Introduction to
Forestry
and
Agroforestry

ABOUT THE AUTHORS

Dr. K.T. Parthiban is borne at the faculty of Forestry and is currently positioned as Dean, Forest College and Research Institute, TNAU, Mettupalayam. He is actively involved in teaching, research and extension activities for the past 25 years. He has conceived the value chain models for Industrial Agroforestry and promoted in association with wood based industries. He is also pioneer in development of varieties in trees. He has handled numerous projects funded by DST, DBT, ICAR, NOVOD Board and several wood based industries. He has published 12 books and over 100 research papers. Some of his books like Objective Forestry, Subjective Forestry to IFS Aspirants and Forestry Technologies have profound utility by the students, professional foresters and wood based industries. Dr.K.T.Parthiban travelled widely in European, South American and Asian countries and contributed towards Forestry Research and Development.

Dr. N. Krishnakumar is borne at the faculty of Forestry and is currently working as a Forest Range Officer (Trainee) in the Tamil Nadu Forest Department. He is specialized in tree breeding and involved in varietal development programme of Thornless bamboos and Sandalwood. He is a recipient of Rajiv Gandhi National Fellowship and has published 3 books and several research papers in the journals of national and international repute.

Dr. M. Karthick is a Scholar borne at the faculty of Forestry and is currently working as Forest Range Officer (Trainee) in the Tamil Nadu Forest Department. He has been involved in Improvement and Conservation of Red Sanders genetic resources through breeding and molecular tools. He is pioneer in improvement of Acrocarpus fraxinifolius and identified clones amenable paper and plywood production.

Introduction to Forestry and Agroforestry

K.T. Parthiban
N. Krishnakumar
M. Karthick

SCIENTIFIC
PUBLISHERS

Published by
SCIENTIFIC PUBLISHERS (INDIA)
5 A, New Pali Road
P.O. Box 91
Jodhpur - 342 001 INDIA

Print 2025

ISBN: 978-93-88172-67-7

Visit the Scientific Publishers (India) website at
http://www.scientificpubonline.com

Printed in India

PREFACE

Forestry as a professional course has started its inception in 1985 under the consideration of Indian Council of Agricultural Research in order to create the professional man power needs of the country. This has attracted several State Agricultural Universities which resulted in the birth of exclusive forestry colleges and the professional B.Sc (Forestry) degree programme across the country. Similarly forestry and agroforestry as a subject of interest has been included in several undergraduate programmes of B.Sc (Agriculture), B.Sc (Horticulture), B.Sc (Sericulture), B.Sc (Botany), B.Sc (Environmental Science). These educational programme demanded comprehensive text books to cater to the educational needs of the graduating scholars. Hence the current book on *Introduction to Forestry and Agroforestry* has been conceived and presented with the - state of the art recent developments in forestry and agroforestry.

The book encompassed 19 chapter which ranges from basic aspects of forestry like Forests and Forestry, Forest Classification and Types, Forest Resources of Important Countries, The State of World's Forest, The State of Indian Forest, to Silviculture of Commercial Important Trees and Silvicultural Principles and Practices. Apart from the fundamental concepts book also incorporated the application oriented topics such as like Forest Utilization, Forest Mensuration, Agroforestry, Social Forestry, Joint forest management and Dendro Energy Resources. These subjects could help the graduating scholars in understanding the practical as well application oriented aspects of forestry. The incorporation of environmental aspects of forestry particularly global warming and carbon sequestration will help the scholars to acquaint the knowledge on environment. The sixteenth chapter on wood based industries shall give overall industrial participation in forestry sector. The inclusion of organizations involved forestry development and the important dates and events in forestry will extent all needed information to the scholars not only for academic purposes but also for all their competitive needs. The last chapter on Acts and Policies will be useful to the graduating scholars on the importance of legal and policy issues in forest management and conservation.

This book on *Introduction to Forestry and Agroforestry* has been designed not only for forestry graduates but also, for other graduates of Agriculture, Horticulture, Sericulture etc., to meet their academic and competitive examinations demands. The book will also act as a master piece of information for all the graduates and other stakeholders to meet the academic requirement besides satisfying the competitive examination requirement.

K.T. Parthiban
N. Krishnakumar
M. Karthick

CONTENTS

CHAPTER - 1

FORESTS AND FORESTRY

Forests are heritage to mankind and provide all basic needs of the society. The word forest is derived from a Latin word *"**Foris**"* which means outside the village boundary and away from the inhabited land. This forest covers wide range of trees, shrubs, herbs, and grasses which are interacting among themselves and create specific ecosystem generally forest have been treated as a productive resource mainly for the production of timber and other forest products. However forestry is a science which deals with the establishment, management and utilization of forests. This chapter deals the basics and fundamental aspects of forest and forestry for use by the graduating professionals and other graduates satisfying academic and other competitive exam needs.

The reader should understand the following objective of the chapter

- Definitions of terminologies
- Historical background of forests
- Objectives of forestry
- Role of forests - Productive, Protective and Ameliorative
- Reasons for declining forest cover

DEFINITIONS OF TERMINOLOGIES

Forest

- "An area set aside for the production of timber and/or other forest produce, or maintained under woody vegetation for certain indirect benefits which it provides, e.g., climatic or protective" (BCFT).
- "Forest is an area of land proclaimed to be a forest under a forest law" (Legal definition).
- "Forest is a plant community predominantly of trees and other woody vegetation, usually with a closed canopy" (Ecology based).

Forestry: "The theory and practice of all that constitutes the creation, conservation and scientific management of forests and utilization of their resources" (BCFT).

Herb: Herb is defined as plant whose stem is always green, tender and height is usually not more than one meter.

Shrub: A woody perennial plant differing from a perennial herb in its persistent and woody stem and less definitely from a tree in its low stature and its habit of branching from the base.

Tree: A large woody perennial plant having a single well defined stem (bole or trunk) and a more or less definite crown.

Crown: The upper branchy part of a tree above the pole.

Crown development: The expansion of crown measured as crown length and crown width.

Stem: The principal axis of plant from which buds and shoots are developed.

Bole: Only lower parts of the stem upto a point where the main branches are given off.

Clean bole:#The part of the bole that is free of branches.

Pole: A young tree from the time when the lower branches begin to fall off to the time when the rate of height growth begins to slow down and crown expansion become marked.

Seedling: Seedling is a plant grown from seed till it attains a height of about one meter.

Sapling: A young tree from the time when it reaches about one meter (3 feet) in height till the lower branches begin fall.

HISTORICAL BACKGROUND OF FORESTS

YEAR	HISTORICAL EVENTS
BC	
2000	• The Agni Purna stated that man should protect trees to have material gains and religious blessings.
1000-700	• Ramayana – Mahabharta – Excellent account of forest; Dandakaranya, Khandav Ban, Nandan Ban
500	• Gautama Buddha preached that man should plant a tree every five years.
300	• Chandra Gupta Maurya realized the importance of the forests and appointed a high officer to look after the forests.

YEAR	HISTORICAL EVENTS
	• Ashoka stated that wild animals and forests should be preserved and protected. He launched programmes to plant trees on a large scale.
321-180	• Mauryan Period - first historical evidence of organized forest administration, planting of trees along road sides and camp sites.
AC	
1500-1600	• The Mughals showed more interest in gardens and their development. Akbar ordered the planting of trees in various parts of his kingdom. Jahangir was well known for laying out beautiful gardens and planting trees.
1790	• Tipu sultan introduces Eucalyptus in Nandi hills, Mysore.
1800	• Commission on Forestry appointed to enquire into the availability of Teak in Malabar region. • Based on felling of Teak below 21 inches is prohibited. Heavy destruction due to European demands.
1806	• Captain Watson appointed as first Conservator of forest by Govt. of Madras.
1823	• Conservator of forest post is abolished.
1828	• Wasteland declared as government properties.
1842	• Conolly, Collector of Malabar region establish Nilambur Teak plantation. First teak plantation raised in Nilambur.
1843	• Afforestation work started. • Champell introduces *Eucalyptus pinnata* at Wellington, Madras
1854	• First Zoo in India (Calcutta Zoo, West Bengal).
1855	• Charter of the Indian Forest- Memorandum is submitted by MC Clelland to Lord Dalhousie.
1855	• British Indian govt issued memorandum outlining rules for forest conservation in the forestry.
1856	• Cleghorn is appointed as first regular Conservator of forest. • Dietrich Brandis appointed as Superintendent of Pegu forest by Lord Dalhouise.
1858	• Introduction of Acacia and Eucalyptus in Nilgiris, Madras.
1864	• Forest department set up. • Scientific forest management started in India.

YEAR	HISTORICAL EVENTS
	• Dietrich Brandis appointed as first Inspector General of forest of India.
1865	• Indian Forest Act passed.
1866	• Changamanga fuel wood plantation established.
1867	• IFS (Indian Forest Service) started.
1878	• First Indian forest act came in to being forests classified as reserve and protected.
1879	• Teak plantation at Makum. • Reorganization of forest service as CF, DCF and ACF posts created.
1870	• Wattles introduced in Ooty and Kodaikanal.
1871	• Working plan started at Nilambur. • Cattle Trespass Act is passed.
1875	• First copy "The Indian Forester" journal issued by Baden Powell and Dr. Schlich.
1878	• Forest school at Dehradun started. • F. Bailey - First Director of the Forest School. • Gamble starts research on Non-Wood Forest Products (NWFP).
1879	• Elephant Preservation Act passed.
1882	• Madras Forest Act passed.
1883	• BNHS (Bombay Natural History Society) started. • Brandis retires and Schlich becomes the second IG of forest.
1884	• Forest school renamed as Imperial Forest School.
1885	• Training for forest officers of India started at Cooper's Hill, England.
1887	• Wild bird protection act enacted.
1890	• Dr. Voelkar from Germany starts study on Indian agricultural problems.
1894	• First Indian Forest policy passed.
1895	• First working plan-Nilambur completed.

YEAR	HISTORICAL EVENTS
1898	• Vedanthangal declared as a first wildlife sanctuary in India.
1901	• Gass Forest museum established in Madras.
1905	• World Forestry Congress set up. • Systematic Research in pulp and paper in India started.
1906	• Imperial Forest School renamed as Imperial Forest Research Institute now it is known as Forest Research Institute (FRI), Dehradun. • Ruther appointed as First Chief Conservator of forest in India.
1909	• Royal Commission on Decentralization set up.
1910	• Board of forestry created at national level.
1912	• Wild birds and Animals Preservation Act passed.
1914	• Destructive Insects & Pests Act passed. • Board of Forestry set up.
1914- 1918	• First world-war heavy destruction due to supply of timber and charcoal to troops.
1919	• Forestry research officers appointed.
1921	• National character of forest administration diluted as forests became provincial subjects.
1927	• Indian Forest Act passed.
1928	• Royal Commission on agriculture in India established.
1929	• Ornithological Survey in Indian conducted.
1932	• Teak stump planting started in Madras.
1934	• New Forest Botanical Garden established.
1935	• Wildlife Birds and Animal (amendment) Act passed.
1936	• First National Park of India named as "Hailey".
1938	• Madras Forest College closed.
1939	• Research in Indian Forest soil stared.
1940-1944	• Second world-war heavy destruction due to demands for defence purposes.
1945	• Madras Forest College reopened.

YEAR	HISTORICAL EVENTS
1947	• Import and Export Control Act passed.
1948	• Central Board of Forestry was set up. • Forest Colleges handed over to Govt. of India.
1949	• M.D. Chaturvedi – First Indian Inspector General of Forest.
1950	• Vanamahotsav- the planting festival is started by K.M. Munshi
1952	• Second Forest Policy passed. • India Board for Wildlife established.
1954	• Minor Forest Products term coined in World Forestry Conference.
1955	• Hill Area (Preservation) Act passed.
1957	• Hailey National Park renamed as Corbett National Park.
1958	• Wildlife Preservation Society of India established.
1959	• Logging Research Institute at Batote established. • Tree Improvement work is started at FRI, Dehradun.
1960	• Prevention of Cruelty to Animals Act passed.
1961	• Forest seed testing started in FRI, Dehradun.
1963	• G.A. Jones conducts aerial survey of Kulu forest.
1965	• Pre-Investment Survey of Forest Resources published the Forest Type Map of India.
1966	• Seeds Act passed.
1968	• Champion and Seth classify the forests of India into 16 forest types. • Westoby coined the word "SOCIAL FORESTRY" in world forestry congress held in India.
1970	• Gujarat- is the first state to start Social forestry programme in India. • Project "Hangul" started in J&K.
1972	• Project Lion started at Gir sanctuary. • Wildlife (Protection) Act passed.
1973	• Project Tiger started. • Chipko movement started by Sunderlal Bahughuna.

YEAR	HISTORICAL EVENTS
	• UNESCO starts MAB (Men and Biosphere) program. • Water (Prevention & Control) Act passed.
1975	• The Crocodile Breeding and Management project launched.
1976	• Forest and Wildlife brought on concurrent list in the 7th schedule by 42"d Amendment to the Constitution. • India becomes the member of CITES. • NCA report was published.
1978	• Central Crocodile Breeding Institute in Hyderabad is established. JFM (Joint Forest Management) started in West Bengal.
1979	• Social Forestry project started.
1980	• Forest (conservation) Act passed.
1981	• Air (prevention & control) Act passed.
1982	• IIFM (Indian Institute of Forest Management) established at Bhopal. • WII (Wildlife Institute of India) started at Dehradun.
1983	• First National Wildlife Action plan adopted.
1984	• MOEF (Ministry of Environment and Forest) created.
1985	• B.Sc. (Forestry) course started in State Agricultural Universities. • NWDB (National Wasteland Development Board) Established.
1986	• Wildlife protection (amendment) act enacted. Trade in animals and in animal articles prohibited. • Environment (Protection) Act passed. • JFM started in Gujarat. • First biosphere reserve in India – Nilgiri Biosphere Reserve.
1987	• FSI (Forest Survey of India) publishes first State of Forest Report. • ICFRE (Indian Council of Forest Research & Education) established at Dehradun. • IGNFA (Indira Gandhi National Forest Academy) established at Dehradun.
1987	• Air (Prevention and Control of Pollution) Act passed.
1988	• Third National Forest Policy passed in India. • New Seed Development Policy passed.

YEAR	HISTORICAL EVENTS
1989	• Biotechnological Centre for Tree Improvement established at Tirupathy.
1991	• Center for Social Forestry and Eco rehabilitation established. • Ibex and Musk deer projects are started.
1991	• Wildlife protection (amendment) act, 1991 passed. Schedule plants were given protection and commercial felling in wildlife sanctuaries banned.
1992	• Project Elephant started.
1992	• National conservation strategy was formulated.
1993	• Great Indian Bustard project started. • Environmental action plan, 1993 formulated.
1995	• Siberian Crane Project started at Bharatpur (Rajasthan).
1996	• Seeds Act passed.
1998	• Norman Borlaug award started. • National Zoo Policy adopted.
1999	• Ministry of Environment and Forests prepares a National Policy. • Macro level Action Strategy on Biodiversity through a consultative process.
2000	• Biodiversity Bill introduced in the parliament Kanchenjunga (Sikkim) declared as the 12th biosphere reserve in the country.
2001	• Protected Area Network Cell of Wildlife Institute of India released the list of National Parks (89 No.) and Wildlife Sanctuaries (486 No.) of India. • S.K. Pandey becomes the Director General of Forest. • Task Force on Greening India.
2002	• PM of India released the National Wildlife Action plan (2002-2016). • Biological Diversity Act passed. • National Afforestation Programme (NAP) was launched. • MoEF launched an Integrated Forest Protection Scheme during 10th Five years plan (2002-2007) to provide holistic support to forest protection like forest fire control management.

YEAR	HISTORICAL EVENTS
	• Wildlife (Protection) Amendment Act.
2003	• Establishment of National Biodiversity Authority at Chennai.
2004	• MoEF notified a Compensatory Afforestation Management and Planning Authority (CAMPA).
2006	• TPP (Twenty Point Programme) started. • National Bamboo Mission started. • FRA (Forest Right Act) passed.
2007	• Wildlife Crime Control Bureau constituted by the Central Govt.
2008	• Implementation of Forest Right Act (FRA).
2010	• National Green Tribunal Act (NGT) passed.
2011	• Green India Mission (GIM) is one of the eight missions under the National Action Plan on Climate Change (NAPCC) was started.
2012	• Land acquisition Bill.
2013	• Criminal Law (Amendment) Act.
2014	• 3rd meeting held on World Congress in Agroforestry, New Delhi. • National Agroforestry Policy enunciated.
2015	• World forest congress held at Durban.
2017	• Bamboo exempted from trees list by amendment in Indian Forest Act 1927.
2018	• Draft national forest policy released for public opinion.

OBJECTIVES OF FORESTRY

The main objectives of forestry are

i. Protect the existing forest and natural resources from all biotic and abiotic factors.

ii. Produce the maximum possible wood and non wood forest produce for domestic and commercial utilisation.

iii. Develop forest and forestry to benefit the society.

iv. Protect the environment and aid in creating Clean Development Mechanisms.

ROLE OF FORESTS

Forests play a significant role in social, cultural, economical, industrial and environmental development of India. Forests are the resource base for sustenance of population and a store house of biodiversity and the associated natural resources. All land use patterns like agriculture, horticulture, animal husbandry etc., either originated or is dependent on forest and forest lands. The role of forests can be primarily classified as productive, protective and other service functions.

Forests play a vital role by extending productive, protective and ameliorative role. The productive role includes supply of wood and non-wood products coupled with meeting other direct demands of the society. The protective role includes the role of forests in protection and conservation of environment, soil, habitat, water, biodiversity, wild animals etc. The Productive, Protective and Ameliorative role are discussed elaborately under the following heads.

A. Productive Role of Forests

i. **Fuel wood:** Wood is used as fuel for thousands of years, until the advent of coal, oil, gas, electricity, etc. Wood constitutes as chief source of fuel. Even today more than half of the total world consumption of wood is for fuelwood. Wood remains the major source of domestic fuel in India. Approximately 330 million m^3 of wood is used as fuel in the country.

   The important fuel wood species are *Acacia* spp., *Casuarina equisetifolia*, *Prosopis*, *Azadirachta indica*, *Leucaena leucocephala*, etc.

ii. **Timber:** Forests act as a resource for supply of timber needs of the society till the recent past. However after the Supreme Court order 1996, only timber from approved working plan areas alone is available. Currently the trees grown outside the forests in the form of agroforestry is able to provide over 70% of the timber demands of the country.

   Major timber species: Teak, Sal, Rosewood, Pterocarpus, Terminalias, Cedar, Gurelina etc.

iii. **Non timber resources:** Forests provide wide range of non timber products like tendu leaves, tamarind, amla, gums, resins, dyes, honey, medicinal plants, grasses, TBOs etc. which provide direct benefit to the society through income generation activity. The NTFP contributes over 60% of the country's forestry export and thus it is one of the major tangible benefits.

iv. **Fodder:** Forests extend fodder to the cattle due to their richness in protein, fiber and fat. The forest in India meet about 30% of the fodder requirement through provision of grazing and browsing facility. It also supplements the fodder needs of the cattle during lean season.

Fodder trees: Acacia, Albizia, Hardwickia, Pterocarpus, Robinia etc.

v. **Industrial raw material:** Forests play a vital role in supplying raw material to the wood based industries like pulp and paper, plywood, veneer, pencil, sports goods, packing, construction industry and agriculture implements. It has been estimated that the country needs over 152 million m^3 of wood as a raw material requirement of various wood based industries and forests are the major resources.

Pulp and paper	: Bamboos, *Eucalyptus, Casuarina*
Plywood	: Eucalyptus, Melia, Toona
Timber	: Teak, Rose wood, Terminalia
Packing cases	: *Pinus* sp., Silver oak, Fir
Matchwood	: Ailanthus, Simaruba, Bombax
Toys	: Adina, Red sanders, Rose wood

vi. **Dendro power:** Forests also contributes towards power generation. It is estimated that 1.5 tonnes of wood is able to generate one MW power which will help to generate decentralized power production and help to resolve the energy crisis.

Example: Prosopis, Casuarina, Eucalyptus, Subabul, Acacia

vii. **Food:** Forests provide a wide range of food for the dwellers and the local tribes and thus contributed towards the food security. The food from forests includes leafy vegetables, mushrooms, berries, fruits and rhizomes which are available in different seasons and which form useful component of the tribal diet. Wild fruits rich in vitamin A and C are also contribute towards food security and nutritional health of forest dwellers.

B. Protective and Ameliorative Role of Forests

i. **Forest and hydrological cycle:** Forests are the vital resources for watersheds. Water and fertile soil are most important pre-requisite for food security. The forests due to their thick humus layer and the soil retention capacity of tree roots are very important towards maintaining and regulating water flows and sub soil water regimes. Almost all the water bodies like rivers are originated predominantly from forest lands and forest derived water tables.

ii. **Forest and ecological security:** Forests are the habitats for diverse species and sustain the ecological diversity. The viable populations of flora and fauna housed within the forests, help to maintain the ecological food chain and maintains the ecosystem sustainably. This will help to protect the flora and fauna.

iii. **Forest and ecotourism:** The concept of ecotourism is gaining significant attraction due to urbanization and population explosion. For this purpose, the forests and other protected areas are acting as a hub for ecotourism activities. Forests indirectly help to create recreational benefits coupled with aesthetic values which are very essential for sustenance of human life.

iv. **Forest and pollution:** Forests are excellent sink for combating various pollutants. They absorb more carbon-dioxide and release oxygen through photosynthetic process and help to balance carbon concentration in the atmosphere. Due to this process, the forests help to reduce the global warming and mitigate acid raining. Forests are one of the best land use systems to mitigate and adapt to growing climate change associated risks.

v. **Forest and climate:** Forests help to moderate the climate due to their multifarious actions. The interception of solar light by the tree canopy helps to augment the micro climate by reducing the temperature and the associated evapo transpiration. The increasing humidity and the increase in precipitation are also essential intangible benefits extended by the forests.

vi. **Forest and soil:** One of the major intangible benefits extended by the forests is its role in soil conservation. The tree intercepts soil and water erosion and improves the soil physical, chemical and biological properties through the addition of leaf litter followed by decomposition. The soil structure is enhanced which lead to increased permeability and arrests soil and water erosion thereby protect the soil.

vii. **Forest as carbon sinks:** Forests and plantation forestry fix enormous quantity of carbon and it is estimated that 1 tonne of wood produced is equivalent to 0.5 tonnes of carbon sequestered. India has a potential of 175 million tonne of carbon to be sequestrated annually through land use and forestry operations.

viii. **Urban forestry:** Rapid urbanization warrants an increased emphasis on urban and recreational forestry. This will include avenues, green belts, parks, gardens and recreational forestry activities. Of late, there has been increasing interest on the role of trees in recreational forestry and salinity the aesthatic needs of the society.

REASONS FOR DECLINING FOREST COVER

Since independence, forests witnessed serious exploitation both for domestic utilization and for other developmental projects which had led to decline in forest cover. The following are the major causative factors for decline of forest.

i. **Population explosion and over exploitation:** The growing population and the associated need for forest and forest products are one of the major reasons for decline in forest area. Since independence the population has been growing at an alarming rate from 300 m to 1200 m population. This population increase has demanded wood for construction, shelter and other forest products which resulted in decline of forest cover. It has been estimated that forest decline in the country was 1.5 m ha/annum till the recent past.

ii. **Developmental projects:** After independence, the developmental projects like river valley projects, dams, roads and communication, railways etc. have taken heavy toll of Indian forests. Examples of Narmada river valley projects, konkan railway project, coal minings etc. are the major cause for declining forest cover.

iii. **Increase in wood demand:** Traditionally forests have been looked for its production potential and it has provided wide range of goods and services both for domestic and industrial utility. The 1952 National Forest Policy has also made a provision for providing raw material resources for the wood based industries. All these issues have led to decline forest cover.

iv. **Diversion forest for agriculture:** The people in general and the tribal people in particular heavily rely on the forest land resources for their livelihood. The political compulsions have led to diversion of forest land for agriculture purpose which acted as a major cause for declining forest cover.

v. **Shifting cultivation:** It is one of age old practice of agroforestry practiced predominantly in North Eastern states and other forest types of North and South India. The ethinic group dwells within the forest have used shifting cultivation process as a source of subsistence and livelihood farming. Through this approach, the local dwellers clear the forest area and practice agriculture to meet their food requirement. The practice will be continued for 2.3 years and based on the productivity the area will be shifted to new area. This type of shifting cultivation across the country is one of the major causes for declining forests in the country.

vi. **Over grazing:** Similar to human population the country has also witnessed higher livestock population. The livestock population growth has not been commensurated with fodder development programme which resulted in shortage of both green and dry fodder. This has compelled the graziers to extensively deploy the livestock in the forested area which is one of the major reasons for decline of forest cover.

vii. **Forest fire:** The fire both natural and intentional is one of the major causes for declining forest cover. The occurrence of natural fire in western ghats and eastern Himalayas have eroded several species. Similarly the smugglers and poachers intensively set fire to collect animal products, charcoal production, honey collection and other forest products collection have led to serious damage and cause the decline of forest cover.

viii. **Extreme climatic conditions:** The decline in forest cover is also attributable due to extreme climatic factors. The increase in temperature due to climate change has created more aridity and create decline in forest cover. Similarly heavy precipitate intensive snow and frost also act as a major causative factors for decline of forest cover.

These are the major causative factors which play a critical role in declining forest cover which resulted in the actual forest cover of around 21.54% compared to the projected cover of 33%.

■■■■

CHAPTER - 2

FOREST CLASSIFICATION AND TYPES

India has mandated establishment of 33% of forest cover against the estimated forest cover of 24.39% as per the 2017 state of report of Forest Survey of India (FSI). Within this 24.39% of forest cover, the actual forest which contributes only 21.54%. This includes wide range of forest types and classes are grouped mainly to treat the forest for various purposes like forest management, forest administration, and utility and also based on growing stock. The forest further has been classified based on climate and vegetation to understand the forest classification and forest type classification to the readers. This chapter briefs various classifications of forest and its types which are covered in the following subheadings.

The reader should able to understand the following objectives of the chapter.

- Forest Types - Definition
- Classification of forests
    - Regeneration
    - Age
    - Composition
    - Ownership
    - Object of management
    - Growing stock
- Classification of forest based on vegetation and climate
- Factors influencing forest vegetation

FOREST TYPES

Forest type is a category of forest defined with reference to its geographical location, climatic and edaphic features, composition and condition.

Champion and Seth defined it as – a unit of vegetation which possesses (broad) characteristics in physiognomy and structure sufficiently pronounced to permit its differentiation from other such units.

CLASSIFICATION OF FOREST

The forests in the country is classified on the basis of

- Method of regeneration
- Age
- Composition
- Objects of management
- Ownership and legal status and
- Growing stock

i. Classification Based on Method of Regeneration

Based on regeneration, the forests are classified as

High forest : Forest regenerated through seeds.

Coppice Forest : Forest regenerated through coppice.

ii. Classification Based on Age

Based on age, the forests are classified as

Even aged forests : Forests consisting of trees which are approximately of the same age (differences allowed up to 25%).

Uneven aged forests : Forests which vary widely in age and contain trees of all ages (the range of age difference being usually more than 20 years).

iii. Classification Based on Composition

Based on composition, the forests are classified into

Pure forest : A forest composed of almost entirely of one species, usually to the extent of not less than 80%. It is also called pure crop or pure stand.

Mixed forest : A forest composed of trees of two or more species intermingled in the same canopy; in practice, and by convention, at least 20% of the canopy must consist of species other than the principal one.

iv. Classification Based on Objects of Management

Based on the objects of management, the forests are classified into

Production forest : It is a forest managed primarily for its produce viz., timber, non-timber and other industrial utility.

Protection forest : It is an area wholly or partly covered with woody growth, managed primarily to regulate stream flow, prevent erosion, hold shifting sand or to exert any other beneficial influence on the physiographic features of the site

Farm forest : It is a forest raised on farms and its adjoining areas either as individual scattered trees or as wood lots to meet the domestic and / or industrial wood requirement.

Recreation forest : It is a forest managed to meet the recreational needs of the urban and rural populations

v. Classification based on ownership and legal status

On the basis of ownership, forests are classified into

State forest : It is a forest owned by state

Community forest : Forest owned and generally managed by a community such as a village, town, tribal authority or local government, the members of which share in the produce or proceeds of sale

Panchayat forest : It is a forest where management is vested with a village panchayat (i.e., a body of men people elected by the villagers from among themselves for specific administrative or other purposes pertaining to the village).

On legal basis, the forests are classified into

Reserved forest : It is an area so constituted under section 3 in chapter II of the Indian Forest Act 1927.

Protected forest : It is an area subject to limited degree of protection under the provisions of Chapter IV of the Indian Forest Act, 1927.

Village forest : A village forest is a state forest assigned to a village community under the provisions of chapter III of the Indian Forest Act, 1927.

vi. Classification on the basis of growing stock

On the basis of growing stock, the forests are classified into

Normal forest : It is a forest with normal growing stock, normal series of age gradation and normal increment.

Abnormal forest : It is a forest with abnormal growing stock, i.e., growing stock in deficit or in excess or in which the relative proportions of the age or size classes are defective resulting in abnormal increment.

CLASSIFICATION OF FOREST BASED ON VEGETATION AND CLIMATE

Systems of Classification of Forest Types

The system of classification of vegetation is based on

i. Botanical, i.e., based mainly on vegetation,

ii. Climatic, i.e., based mainly on climate or

iii. Ecological, i.e., based mainly on ecosystem consisting of vegetation environment complex.

Champion and Seth revised the preliminary classification of forest types of India prepared in 1935. In this classification, the forests of India have first been divided into the following five major groups

i. Tropical forests

ii. Montane subtropical forests

iii. Montane temperate forests

iv. Sub-alpine forests and

v. Alpine scrub

The five major groups are divided into 16 type groups or, simply, groups on the basis climatic data and vegetation. The tropical forests have been divided into seven groups, the montane subtropical forests into three groups, montane temperate forests into three groups, sub-alpine forests into one group and alpine scrub into two groups. Most of these groups or type groups have been further divided into two subgroups like southern and northern forms.

FOREST TYPES: DISTRIBUTION AND DESCRIPTION

Sl.No.	Forest types	Distribution	Description
1.	Tropical Wet Evergreen Forests	Tamil Nadu, Kerala, Karnataka, Maharashtra, Andaman and Nicobar Islands, North-eastern parts particularly in West Bengal, Assam and Orissa.	This type has Dense tall forest, entirely evergreen or nearly so. This type of forest has two sub-groups i.e., Northern tropical wet Evergreen forests and Southern tropical wet evergreen forest
2.	Tropical Semi-Evergreen Forests	Tamil Nadu, Kerala, Karnataka, Maharashtra, Goa, Gujarat, Andaman, Assam, West Bengal and coastal parts of Orissa.	Dominants are predominantly evergreen and also includes deciduous. It has two sub groups i.e., Sothern tropical semi-evergreen forest and Northern tropical semi-evergreen forest.

Sl.No.	Forest types	Distribution	Description
3.	Tropical Moist Deciduous Forests	Tamil Nadu, Kerala, Karnataka, Maharashtra, Gujarat, Madhya Pradesh, Uttar Pradesh, Bihar, Orissa, Bengal, Assam and Andaman and Nicobar Islands.	Dominants mainly deciduous but some dominants and lower storey largely evergreen. Top canopy rarely dense and even but over 25 m high. This type has three sub-groups i.e., Andaman moist deciduous forest, South Indian Tropical Moist Deciduous Forest and North Indian Tropical Moist Deciduous Forest.
4.	Littoral and Swamp Forests	Coastal region, Estuaries of rivers, Along Ganga and Brahmaputra rivers, Along the banks of the largest streams in hilly tracts, Himalayan sub-montane forest, Kerala.	Mainly evergreen of varying density and height but always associated predominantly with wetness. It has four sub-groups. i.e., Littoral Forest, Tidal swamp Forest, Tropical Fresh water swamp Forest, Tropical seasonal swamp Forest, Tropical Riparian Fringing Forests.
5.	Tropical Dry Deciduous Forests	Tamil Nadu, Karnataka, Andhra Pradesh, Maharashtra, Gujarat, Rajasthan, Uttar Pradesh, Madhya Pradesh, Bihar, Orissa, Himachal Pradesh, West Bengal.	Entirely deciduous or nearly so. Top canopy rather light and rarely over 2m height, usually 8 to 20m. *Sub types:* Southern tropical dry deciduous forest Northern tropical dry deciduous forest
6.	Tropical Thorn Forests	Gujarat, Rajasthan, Uttar Pradesh, Madhya Pradesh, Semi-arid areas of Punjab, Dry peninsular India.	Deciduous with low thorny trees and xerophytes predominating. Canopy more or less broken. Height under 10m. *Sub- types:* Southern tropical thorn forest Northern tropical thorn forest
7.	Tropical Dry Evergreen Forests	It is distributed along the coastal belt of Tirunelveli to Nellore	Hard leaved evergreen trees predominate with some deciduous emergent often dense but under 20m height.

Sl.No.	Forest types	Distribution	Description
8.	Subtropical Broad leaved Hill Forest	Southern part of Indian Hills Eastern hills and Khasi hills	Broad leaved largely evergreen *Sub types:* Southern Subtropical Broadleaved Hill Forest Northern Subtropical Broadleaved Hill Forest
9.	Subtropical Pine Forests	Hills of central Himalayas, Nagar and Manipur hills	Pine association predominates
10.	Sub-tropical Dry Evergreen Forest	Regions of foothills of western Himalayas, and in Bhabar and Siwalik hills.	Low xerophytic forests and scrubs. The species are tending to form pure crop. The degraded stage of these forests occur in the form of Dodonaea scrub.
11.	Montane Wet Temperate Forests	Hills of Tamil Nadu and Kerala and in the Eastern Himalayas	Evergreen forests without conifers
12.	Himalayan Moist Temperate Forest	Regions of Himalayas	Evergreen mainly temperate sclerophyllous oaks and conifers
13.	Himalayan Dry Temperate Forests	Regions of Kashmir, Uttar Pradesh, Himachal Pradesh, Sikkim and Arunachal Pradesh	Open coniferous forests with sparse temperate xerophytic under growth
14.	Sub-Alpine Forests	Regions of Himalayas	Stunted deciduous or ever-green forests usually in close formation, forest with or without conifers
15.	Moist Alpine Scrub	Areas above 3400m altitude	Low but often dense scrub
16.	Dry Alpine Scrub	Areas under Xerophytic condition	Xerophytic scrubs in open scrub formation.

FOREST TYPE: CLIMATIC REQUIREMENT AND SPECIES COMPOSITION

Sl.No.	Forest types	Climatic Requirement: Mean Annual Temperature	Climatic Requirement: Mean Annual Rainfall	Species Composition
1	Tropical Wet Evergreen Forests	22- 27ºC	2000-6400 mm	*Olea dioica* *Schleichera trijuga* *Knema attennata* Hopea species *Syzygium cumini* *Diospyros assimilis* *Terminalia paniculata* *Hotigarna arnottiana* *Linociera malabarica* *Vateria indica*
2	Tropical Semi-Evergreen Forests	24- 26ºC	1500-3000mm	*Terminalia paniculata* *Castanopsis species* *Terminalia crenulata* *Tectona grandis* *Schima wallichii* *Olea dioica* Macaranga species *Xylia xylocarpa* *Syzygium cumini* Albizia species
3	Tropical Moist Deciduous Forests	21- 27ºC	1000-3000mm	*Tectona grandis* *Shorea robusta* *Terminalia crenulata* *Anogeissus latifolia* *Cleistanthus collinus* *Lannea coromandelica* *Terminalia paniculata* *Lagerstroemia parviflora* *Xylia xylocarpa* *Mallotus philippinensis*

Sl.No.	Forest types	Climatic Requirement		Species Composition
		Mean Annual Temperature	**Mean Annual Rainfall**	
4	Littoral and Swamp Forests	26- 29°C	760-5000mm	*Avicennia marina* *Excaecaria agallocha* *Avicennia officinalis* *Aegiceras alba* *Dalbergia sissoo* *Trewia nudiflora* *Tectona grandis* *Schleichera trijuga* *Bridelia retusa* *Shorea robusta*
5	Tropical Dry Deciduous Forests	22- 27°C	2000-6400mm	*Tectona grandis* *Shorea robusta* *Anogeissus latifolia* *Terminalia crenulata* *Lannea coromandelica* *Lagerstroemia parviflora* *Chloroxylon swietenia* *Butea monosperma* *Cleistanthus collinus* *Diospyros melanoxylon*
6	Tropical Thorn Forests	24- 28°C	460-1250mm	*Anogeissus latifolia* *Albizia amara* *Chloroxylon swietenia* *Hardwickia binata* *Canthium decoccum* *Wrightia tinctoria* *Lannea coromandelica* *Pterocarpus santalinus* Eucalyptus species Grewia species *Anogeissus latifolia*
7	Tropical Dry Evergreen Forests	28- 29°C	870-1270mm	*Albizia amara* *Syzygium mentanum*

Sl.No.	Forest types	Climatic Requirement		Species Composition
		Mean Annual Temperature	Mean Annual Rainfall	
				Anogeissus latifolia *Premna tomentosa* *Canthium decoccum* *Dalbergia paniculata* *Acacia lenticularis* *Chloroxylon swietenia* Albizia species *Memecylon angustifolium*
8	Subtropical Broadleaved Hill Forest	16-22ºC	1500-11400 mm	*Schima wallichii* Castanopsis species *Pinus Kesiya* *Quercus species* *Alnus nepalensis* *Macaranga species* Albizia species *Callicarpa arborea* *Chickrassia velutina* *Acacia auriculiformis* *Schima wallichii*
9	Subtropical Pine Forests	15-22ºC	1000-3000 mm	*Pinus roxburghii* *Quercus spp* *Rhododendron arboreum* *Lyonia ovalifolia* *Mallotus philippinensis* *Schima wallichii* *Shorea robusta* *Myrica sapida* *Acacia catechu* *Quercus semicarpifolia*
10	Sub-tropical Dry Evergreen Forest			*Olea cuspidata* *Acacia modesta* Dodonea scrub

Sl.No.	Forest types	Climatic Requirement		Species Composition
		Mean Annual Temperature	Mean Annual Rainfall	
11	Montane Wet Temperate Forests	12-17°C	1300-6000 mm	*Alnus nepalensis* Castanopsis species Quercus species *Symplocos laurina* Michelia species *Callicarpa arborea* *Symplocos theaefolia* *Abies pindrow* *Schima wallichii* *Canarium bengalense*
12	Himalayan Moist Temperate Forest	13-16°C	1500-3000mm	*Quercus leucotrichophora* *Pinus roxburghii* *Rhododendron arboreum* *Lyonia ovalifolia* *Pinus excelsa* *Quercus semicarpifolia* *Cedrus deodara* *Quercus dilatata* *Abies pindrow* *Abies smithiana*
13	Himalayan Dry Temperate Forests	6-17°C	80-800mm	*Quercus semecarpifolia* *Pinus excelsa* *Abies pindrow* *Cedrus deodara* *Abies smithiana* *Quercus dilatata* *Betula utilis* *Taxus baccata* Acer species *Prunus cornata*
14	Sub-Alpine Forests	2-6°C	10-55 mm	*Abies densa* *Abies pindrow* *Pinus excelsa*

Sl.No.	Forest types	Climatic Requirement		Species Composition
		Mean Annual Temperature	**Mean Annual Rainfall**	
				Quercus semecarpifolia
				Quercus species
				Rhododendron arboreum
				Acer species
				Prunus cornata
				Alnus nepalensis
				Abies smithiana
15	Moist Alpine Scrubs			*Pinus excelsa*
				Quercus species
				Rhododendron arboreum
				Alnus nepalensis
				Phoebe species
				Quercus semecarpifolia
				Rhododendron species
				Abies densa
				Quercus grifpthii
				Abies pindrow
16	Dry Alpine Scrubs			*Juniperus communis*
				Juniperus wallichiana
				Artemisia

FACTORS INFLUENCING FOREST VEGETATION

Vegetation in every place is not unique due to different factors prevailing in different regions. Besides, every plant has its unique requirement for its favourable growth and reproduction. Hence the local factors of the site determine the type of vegetation.

The **factors of locality** are defined as the effective climatic, edaphic, topographic and biotic conditions of a site which influence the vegetation of the site.

Knowledge on factors of locality is essential in order to carry out many forestry practices. It determines the species suitability, adaptability and distribution in any area.

The following factors of locality influence the forest vegetation

S.No.	Major factors of locality	Various components
1.	Climatic factors	a. Light b. Temperature c. Moisture d. Snow e. Wind f. Frost
2.	Edaphic factors	a. Soil physical properties b. Soil chemical properties c. Soil biotic properties
3.	Topographic factors	a. Slope b. Aspect c. Exposure d. Altitude e. Configuration
4.	Biotic Factors	a. Plant based influences b. Animal based influences c. Insect based influences d. Human influences

Influence of Climatic Factors on Vegetation

Light

Light is one of the foremost and essential climatic factors and plays a vital role in influencing forest vegetation due to the following activities.

- Light is very essential for photosynthetic activity and involved in growth and development function
- Light is important for chlorophyll formation.
- Light affects respiration
- The intensity of light is very important and increased intensity has marked influence on the germination and growth of seedlings.
- The photoperiodicity of light affects the growth, seed germination, leaf fall, flowering, fruiting and other growth and developmental activities.
- Light also affects the form and development of trees.

- The plant stratification also influenced by the light and decides the presence of dominant, co-dominant and suppressed trees.
- Based on light requirement, the trees are classified into following types

S.No	Type	Species
1.	Light demanders	*Tectona grandis, Shorea robusta, Terminalia* sp., *Adena* sp., *Dipterocarpus* sp., *Pterocarpus* sp.
2.	Shade bearers	*Artocarpus* sp., *Gmelina arborea, Pterocarpus marsupium*
3.	Shade demanders	*Mesua ferea, Xylia xylocarpa, Schleichera oleosa, Syzygium cumini*

Temperature

- Temperature refers to the relative hotness or coldness and influence the vegetation in several ways.
- Temperature is a source of energy for various metabolic activities like photosynthesis, respiration, transpiration and in turn influences the growth and development of plants.
- Temperature also augments the activities of microbial population in the soil and accelerates the decomposition process.
- Temperature influences all the enzymatic activities and is involved in all metabolic process.
- Temperature influences the growth and development of plants and optimum temperature requirement for plants lies between 25°C and 30°C.
- Higher temperature of more than 45°C cause adverse effect on vegetation. It causes wilting of seedlings due to coagulation of protoplasm while trees usually sustain higher temperatures.
- A phenomenon called **Sun scald** is known to occur at high temperature which leads to the death of cambium in trees.

Temperature influence on forest types

S.No	Forest type	Temperature range (°C)
1.	Tropical	20 to 45
2.	Sub-tropical	20 to 40
3.	Temperate	5 to 30
4.	Alpine	-5 to 20

Influence of lower temperature

The lower temperature leads to occurrence of frost which is nothing but chilling of air below the freezing point. Frost is classified into 3 types

S.No	Type of frost	Characteristics and influences
1.	Radiation frost	a. This frost occurs during winter nights when the sky is clear. b. The temperature near the ground level falls below the freezing point which leads to the formation of ice crystals on soil surface and vegetation. c. It leads to killing of young seedling and young regeneration. d. Radiation frost is otherwise known as ground frost.
2.	Pool frost	a. It occurs due to accumulation of heavy cold air in the valleys and cause heavy injuries and even kills large trees. b. The frost becomes heavy and moves down due to cooling of air and results in convection currents. c. Pool frost is otherwise called as convection frost
3.	Adjective frost	a. This is caused due to movement of cold air from outside particularly during winter season. b. It occurs predominatly

Injuries due to frost

- Young plants and seedlings are seriously affected due to frost. Frost causes freezing of water in the tissues and arrest the growth and development due to accumulation salts, coagulation and precipitation of protoplasm in the cells.
- Due to frost the transpiration gets affected and plants may die.
- The freezing of soil moisture arrest root penetration and may cause dislocation and death.
- The frost also create crack in trees and affects wood quality.
- Frost may also cause canker and influences strength and mechanical properties.
- Trees are classified into frost resistant and susceptible species.

S.No	Parameter	Species suitability
1.	Frost resistant trees	*Eucalyptus,* Dalbergia sp., Toona sp., *Anogeisus latifolia, Acacia catechu*
2.	Frost susceptible trees	Teak, Terminalia, *Boswellia serrata*

3. Water

Water is essential for all physiological process like photosynthesis, respiration, transpiration and influences growth and development.

- Water plays an important role in maintaining the turgidity and growth of plants.
- Water helps to dissolve the nutrients in the soil and make them available to plants.
- The translocation of nutrients within the plant system is facilitated by water.
- It also helps in seed germination, seed dispersal and seedling establi-shment.
- Water helps in soil weathering.

Influence of water on vegetation

The water in the form of moisture is one of the most essential factor in the vegetation development. In India, the water and moisture availability depends on seasonal rainfall which influences the forest type, species composition, growth and development.

4. Wind

Wind is another important climatic factor which affects the growth and form of vegetation.

Influence of wind on vegetation

- Higher wind velocity results in uprooting of trees and cause mechanical injury.
- It also causes excessive evapotranspiration and dries up the plants and soil rapidly.
- The form of the tree and crown is influenced mostly by wind velocity.
- It also leads to soil erosion
- Wind helps in seed dispersal

Bioclimate

Bioclimate is the climate defined by the modified or adjusted climatic factors. The modified climatic conditions produced by the vegetation which differs from the general climate is called bioclimate.

Microclimate

Microclimate is defined as the climate of small areas, which for some reason, differs significantly from the general climate of the area more particularly the climate under a plant or other cover differing in extremes of temperature and moisture etc., form the climate outside that cover.

Importance of microclimate

- Microclimate of different aspects of hills is different which decides regeneration of species. In temperate Himalayas, fir and spruce regenerates well in northern aspect whereas deodar prefers southern aspect.
- It helps in regeneration planning.
- At higher altitude of subtropical zone, northern aspects become too cold for chir which is replaced by kail in nature. Sowings of chir in such places are not likely to be successful.
- In valleys, frost is more common and hence clear felling is avoided.
- Introduction of exotic species needs detailed study of microclimate. Otherwise introduction of exotic become failure.

Edaphic Factor

Edaphic factors predominantly deals with soil structure, composition, texture, etc and its functions. The soil influences the vegetation type through its physical, chemical and biological properties.

The different properties of the soil that influences the vegetation are described below:

Physical properties

i. **Soil texture:** Texture refers the relative proportion of sand, silt and clay in a soil. Some trees prefer sandy soil (*Thespesia populnea, Casuarina, Anacardium occidentale)* whereas some trees grow well only in clay soils (*Acacia nilotica, Prosopis juliflora).*

ii. **Structure:** Structure refers the arrangement of soil particles. The arrangement of soil particles determines the porosity of the soil thus it influences the water retention which in turn indirectly influences the vegetation.

iii. **Surface crusting and hardpan:** Surface crust refers the hard mass of soil formed on the surface of soils having high silt and clay. Hardpan refers the cemented high density soil layer below the surface layer due to the accumulation of clay in sub surface. Surface crust reduces the germination whereas hard pan restricts the root growth.

iv. **Soil depth and stoniness:** The relative depth is also a factor for distribution of plants. Different trees needs different soil depth and stoniness.

v. **Soil moisture:** Soil moisture availability determines the vegetation directly. Humid soil sustains evergreen vegetation and dry soil supports xerophytes.

Chemical properties

i. **Soil reaction:** Soil reaction refers the acidity or alkalinity of the soil. Plant nutrients availability differs with soil reaction. Similarly, certain plants grow well in acid soil (*Xylia xylocarpa, Dipeterocarpus* spp) whereas certain plants prefer alkaline soils (*Tamarix* spp, *Salvadora persica*). In general, most of the plants yield well under neutral soil conditions.

ii. **Electrical conductivity:** Electrical conductivity of the soil indicates the relative soluble salts presence in the soil. Only saline tolerant species (*Salvadora persica, Tamarix* spp, *Prosopis juliflora*) survive well in saline soils whereas other plants prefers neutral soils.

iii. **Calcareousness:** The soil calcareousness determines the electrical conductivity and availability of plant nutrients. Thus it differentiates the plant distribution in calcareous and non-calcareous soils.

Examples of influence of soil properties on plant distribution

S.No	Tree Species	Soil type
1.	Chir pine	Quartzite rocks
2.	Blue pine	Mica schist
3.	Deodar	Alluvium soils
4.	*Cupressus torulosa* and *Ixora parviflora*	Lime stone soils
5.	Teak	Lime rich soils
6.	*Xylia xylocarpa* and *Cleistanthus collimis*	Laterite
7.	*Pteroccirpus santalinus*	Quartzite
8.	*Acacia leucophloea* and *Prosopis spicigera*	Clay soil
9.	*Prosopis juliflora, Acacia arabica, Tamarix aphylla, Salvadora oleoides, Salvadora persica,* and *Sporolobus marginatus*	Saline

Topographic Factors

Topography describes the physical features of a place. Topographic factors are the factors pertaining to the configuration of land surface *viz.,* altitude, slope, aspect and exposure. Any place can be described by its topographic elements such as configuration of land surface, altitude, slope, aspect and exposure and are described hereunder.

i. **Configuration of land surface** denotes the surface configuration or outer shape of the land. It determines the microclimate such as shade, temperature, etc. and thus influences vegetation.

ii. **Altitude** of the place refers its elevation from the mean sea level. Altitude influences solar radiation, temperature, air density and rain-fall. With increasing altitude, the air temperature decreases whereas solar radiation increases. Air density decreases with increasing altitude.

iii. **Slope** refers the slanting nature of surface *i.e.,* one end of the land is at higher elevation and other end at lower elevation. It influences drain-age and runoff caused by rainfall. In general, steep slope increases the runoff and soil erosion.

iv. **Aspect** refers the direction towards which slopes faces. In India, northern and eastern slope of the mountain have relatively low air temperature whereas western and southern slope experience slightly warm temperature.

v. **Exposure** refers the relation of site to weather conditions, especially sun and wind. The site exposed well to sun light has good weathered soil compared to soils found in polar regions which are less exposed to sun light. Also the exposure determines the soil moisture and its availability. All these factors has direct relation to plant growth.

Biotic Factors

Biotic factor refers to the different influence of living organisms on forests.

Influence of plants

At a particular site, there are many numbers of plants. Hence each will influence one another in any of the following ways:

i. **Competition** refers the struggle for limited available food between the same species. As a general rule, survival of the fittest rules the habitat.

ii. **Parasites** are the plants which derive the nutrient on other plants and thereby decrease the vigour of the host plant. Under extreme conditions, the host plant become died.

iii. **Epiphytes** are the plants which live on other plants only for shelter and not for food. It generally weakens the host and does not kill the host.

iv. **Climbers** are plants which depend on other plants mainly for physical support. It reduces the diameter increment of growing trees. The tendrils of certain climbers make a hole on wood which reduces its value and usability.

v. **Weeds** are unwanted plants for that area weeds are plants out of place. It competes with the main species for light, water, space and nutrients. It results in poor growth of main species.

vi. **Fungi** cause many diseases to tree species. In case of mychorrhizae fungus, it augments the phosphorous nutrition to plant through solubilization and mobilization of soil phosphorous.

Influence of insects

Insects feeds plant parts such as leaf, fruit, seed, flower, etc. as their food. Severe defoliation decreases the plant photosynthesis and reduces the plant growth. Feeding of flowers and seeds by insects influences the regeneration negatively.

Influence of wild animals

Wild animals mainly herbivores feeds on the living plants for their livelihood. Thus population of palatable plant species depends on the herbivores population. Excess herbivores decreases the palatable species and *vice versa.* Bear strips the bark of certain trees for its food source. Antlers rub theirs antlers against the tree to shed it off. This activity exposes the bark which enhances the disease infection. Elephants breaks and tramples the young seedlings. Thus wild animals influence the vegetation in an area.

Influence of Man and his animals

Man's influence on the local vegetation is immense. He utilizes the plant for his needs such as food, fuel, wood, shelter, medicine and others. He allows only species that has a more utility. In addition to this, the animals owned by him cause many irrecoverable damages to the ecosystem.

■■■■

CHAPTER - 3

THE STATE OF WORLD'S FOREST

Globally the forests are gaining significant attraction due to the growing demand for wood and wood products besides its role in climate change mitigation and adaptation. The world's forest resources are continue to decline due to anthropogenic pressures coupled with industrialization and associated developmental processes. Hence it is very essential to know the state of global forest resources to understand the developments in forestry sector across the world.

This chapter summarizes the state of world forest in the following sub headings

- Geographical distribution of world forest
    - World forest – Facts
    - Region wise forest area
    - Country wise forest cover
- Classification of world forest
- Productivity potential of world forest
- Increment of world forest

GEOGRAPHICAL DISTRIBUTION OF WORLD FOREST

The forest resources of the world and their management have witnessed paradigm shift during the last two to three decades. The world forest resources continue to decline due to accelerated population growth and the associated development projects. It was estimated that during 1990 the world had 4128 million ha of forest which has come down to 3999 million ha in the 2015 assessment. This decrease in forest cover in terms of percentage has been estimated from 31.6% in 1990 to 30.6% in 2015. The net forest loss during the period between 1990 and 2015 has been estimated at 129 million ha and registered an annual loss of 7.6 million ha.

During the last two years the area under planted forest has increased over 1100 million ha and accounts for 7% of the world forest area. The area under planted forest has increased at the rate of 3.6 million ha during the period between 1990 and 2000 which has increased to 5.2 million ha per year during the period between 2000 and 2010 and it has slow down to 3.1 million ha during the period between 2010 and 2015.

The reason for the decrease was recorded due to decreased planting activities in East Asia, North America, Europe, South and South East Asia. Not only the forest area both natural and planted forest has witnessed changes, but also the production and consumption pattern of wood across the world changes rapidly.

The annual wood removals during 1990 were 2.8 billion m^3 in which fuel wood removal alone recorded 41%. In the year 2011 the annual wood removal increased to 3 billion m^3 almost 30% of the world forest was designated as production forest due to its provisions for multiple uses like timber, non wood forest produce, water, recreation and other values. The forest area based on its service roles along with various disturbances are furnished in Table 1.

Table 1. World Forest – Facts

S.No	Variables	Area
FOREST AREA AND CHARACTERISTICS		
1.	Forest area	3999 million ha
2.	Area of other wooded land	1204 million ha
3.	Area of other land with tree cover	284 million ha
4.	Average annual reforestation	27 million ha
5.	Natural forest	3695 million ha
6.	Planted forest	291 million ha
7.	Net annual forest change	-3.3 million ha
8.	Net annual natural forest change	-6.5 million ha
9.	Net annual planted forest change	3.3 million ha
PRODUCTION		
10.	Forest growing stock	531 billion m^3
11.	Forest growing stock	129 m^3 per ha
12.	Carbon in above- and below-ground biomass	296 Gt
13.	Carbon in above- and below-ground biomass	73 tonnes per ha
14.	Production forest	1187 million ha
15.	Multiple-use forest	1049 million ha
16.	Total wood removal	2997 million m^3
PROTECTIVE FUNCTIONS AND SELECTIVE ECOSYSTEM SERVICES		
17.	Protection of soil and water	1015 million ha
18.	Ecosystem services, cultural or spiritual values	1163 million ha

S.No	Variables	Area
BIODIVERSITY AND CONSERVATION		
19.	Conservation of biodiversity	524 million ha
20.	Primary forest	1277 million ha
21.	Forest area within protected areas	651 million ha
DISTURBANCES		
22.	Area with invasive woody species	79 million ha
23.	Forest area burned	65 million ha
24.	Forest area with reduction in canopy cover	185 million ha

(FAO, 2015)

Region Wise Forest Status in the Word

The world forests have been presented region wise representing Africa, Asia, Europe, North and Central America, Oceania and South American countries. Among the regions, Europe has recorded higher forest area which is estimated at 1110 m.ha and the oceanic countries recorded lower forest area of 174 m. ha. In terms of growing stock the region varied significantly and the forest in oceanic countries registered the growing stock of 202 m^3/ha and Asia recorded lower growing stock of 93 m^3/ha.

The assessment of wood removals indicated that Asian countries recorded increased wood removals of 780 m.m^3 followed by European countries (681 m. m^3). The removal from oceanic countries recorded the lower value of 63 m. m^3. In case of estimations on forest area burnt, South American countries recorded the highest status of 36 m. m^3 followed by African countries (19 m. m^3). The region wise information on forest area including natural and planted forest, productivity, carbon status, ecosystem services are furnished in Table 2.

Table 2. Region wise Forest Status in the World

S.No	Variable	Area					
		Africa	Asia	Europe	North & central America	Ocean-ania	South America
1	Forest area (million ha)	624	593	1015	751	174	842
2	Natural forest (million ha)	600	462	929	707	169	827
3	Planted forest (million ha)	16	129	83	43	4.4	15
4	Net annual forest change (million ha)	-2.8	0.8	0.4	0.1	0.3	-2.0

S.No	Variable	Area					
		Africa	Asia	Europe	North & central America	Ocean-ania	South America
5	Net annual natural forest change (million ha)	-3.1	-1.0	0.01	-0.4	0.3	-2.2
6	Net annual planted forest change (million ha)	0.2	1.8	0.4	0.5	0.03	0.4
7	Forest growing stock (billion m^3)	79	55	115	96	35	150
8	Forest growing stock (m^3 per ha)	128	93	113	129	202	178
9	Carbon in above- and below-ground biomass (Gt)	60	36	45	36	16	103
10	Carbon in above- and below-ground biomass (tonnes per ha)	96	61	45	58	92	122
11	Production forest (million ha)	165	247	511	124	13	127
12	Multiple-use forest (million ha)	133	129	238	391	54	104
13	Total wood removals (million m^3)	614	780	681	513	63	346
14	Protection of soil and water (million ha)	50	195	123	534	37	76
15	Ecosystem services, cultural or spiritual values (million ha)	67	43	122	642	123	167
16	Conservation of biodiversity (million ha)	92	86	53	127	36	130
17	Primary forest (million ha	135	117	277	320	27	400
18	Forest area within protected areas (million ha)	101	115	46	75	27	287
19	Forest area burned (million ha)	19	1.4	2.6	3.9	2.6	36

(FAO, 2015)

Country Wise Forest Cover

The Food and Agriculture Organization (FAO) monitoring and assessment report of 2015 has recorded forest area for almost all the countries in the world. The data indicated in the FAO report has been presented in this book for use by the academicians, researchers, professionals, students and other stake holders. The forest area of 221 countries representing the entire regions of the world is furnished in Table 3. Among these 221 countries, the following 5 countries recorded increased forest area.

S.No.	Country	Forest Area
1.	Russia	814.931 million ha
2.	Brazil	493.538 million ha
3.	Canada	347.069 million ha
4.	USA	310.09 million ha
5.	China	208.321 million ha

All these countries are developed or high income countries. India ranked 10th position in terms of forest area recording 70.682 million ha as per the FAO forest monitoring and assessment report. The table indicates that the forest area is increased income countries registered increasing trend compared to under developed and undeveloped countries. The countries like Anguilla, Bahrain, Bermuda, British virgin Island, Djibouti, Isle of man, Jercy, Kuwait, Liechte-nstein, Maldives, Mayotte, Montserrat, Netherland Antilles, occupied Palestine territory, Pitcairn, Saint Helena, Saint Martin, Tonga, Tuvalu, Wallis and Futuna Islands have recorded a lowest forest area of less than 10,000 hectares.

Table 3. Country Wise Forest Cover

S.No	Country/Territory	Forest		Other wooded land	
		1000 ha	% of land area	1000 ha	% of land area
1.	Afghanistan	1350	2.1	29471	45.2
2.	Albania	772	28.2	256	9.4
3.	Algeria	1956	0.8	2569	1.1
4.	American Samoa	18	87.7		
5.	Andorra	16	34.0		
6.	Angola	57856	46.4		
7.	Anguilla	6	61.1		
8.	Antigua and Barbuda	10	22.3	16	35.5
9.	Argentina	27112	9.9	64376	23.5
10.	Armenia	332	11.8	63	2.2
11.	Aruba				

S.No	Country/Territory	Forest		Other wooded land	
		1000 ha	% of land area	1000 ha	% of land area
12.	Australia	124751	16.2	250961	32.7
13.	Austria	3869	46.9	153	1.9
14.	Azerbaijan	1139	13.8	0	0.0
15.	Bahamas	515	51.4	36	3.6
16.	Bahrain	1	0.8		
17.	Bangladesh	1429	11.0	294	2.3
18.	Barbados	6	14.7	1	2.6
19.	Belarus	8634	41.6	540	2.6
20.	Belgium	683	22.6	36	1.2
21.	Belize	1366	59.9	200	8.8
22.	Benin	4311	39.0	2889	26.1
23.	Bermuda	1	20.0		
24.	Bhutan	2755	72.3	432	11.3
25.	Bolivia (Plurinational State of)	54764	50.6	2473	2.3
26.	Bosnia and Herzegovina	2185	42.8	549	10.8
27.	Botswana	10840	19.1	34791	61.4
28.	Brazil	493538	59.0	39535	4.7
29.	British Virgin Islands	4	24.1	2	11.2
30.	Brunei Darussalam	380	72.1	50	9.5
31.	Bulgaria	3823	35.2	22	0.2
32.	Burkina Faso	5350	19.6	4795	17.5
33.	Burundi	276	10.7	201	7.8
34.	Cabo Verde	90	22.3		
35.	Cambodia	9457	53.6	133	0.8
36.	Cameroon	18816	39.8	12715	26.9
37.	Canada	347069	38.2	40866	4.5
38.	Cayman Islands	13	52.9		
39.	Central African Republic	22170	35.6	32731	52.5
40.	Chad	4875	3.9	19563	15.5
41.	Chile	17735	23.9	14645	19.7

S.No	Country/Territory	Forest		Other wooded land	
		1000 ha	% of land area	1000 ha	% of land area
42.	China	208321	22.1	103637	11.0
43.	Colombia	58502	52.7	16352	14.7
44.	Comoros	37	19.9	26	14.0
45.	Congo	22334	65.4	10479	30.7
46.	Cook Islands	15	64.0		
47.	Costa Rica	2756	54.0	8	0.2
48.	Cote d'Ivoire	10401	32.7	2554	8.0
49.	Croatia	1922	34.3	569	10.2
50.	Cuba	3200	30.1	332	3.1
51.	Cyprus	173	18.7	213	23.1
52.	Czech Republic	2667	34.5		
53.	Democratic People's Republic of Korea	5031	41.8		
54.	Democratic Republic of the Congo	152578	67.3	11513	5.1
55.	Denmark	612	14.4	46	1.1
56.	Djibouti	6	0.2	220	9.5
57.	Dominica	43	57.8		
58.	Dominican Republic	1983	41.0	302	6.3
59.	Ecuador	12548	50.5		
60.	Egypt	73	0.1		
61.	El Salvador	265	12.8	204	9.8
62.	Equatorial Guinea	1568	55.9	9	0.3
63.	Eritrea	1510	15.0	7132	70.6
64.	Estonia	2232	52.7	224	5.3
65.	Ethiopia	12499	11.4	40631	37.1
66.	Fiji	1017	55.7	97	5.3
67.	Finland	22218	73.1	801	2.6
68.	France	16989	31.0	590	1.1
69.	French Guiana	8130	98.6		

S.No	Country/Territory	Forest		Other wooded land	
		1000 ha	% of land area	1000 ha	% of land area
70.	French Polynesia	155	42.3		
71.	Gabon	23000	89.3		
72.	Gambia	488	48.8	99	9.9
73.	Georgia	2822	40.6	7	0.1
74.	Germany	11419	32.8		
75.	Ghana	9337	41.0		
76.	Greece	4054	31.5	2492	19.3
77.	Grenada	17	50.0	1	3.6
78.	Guadeloupe	71	42.5	12	7.0
79.	Guam	25	46.3		
80.	Guatemala	3540	33.0	1334	12.4
81.	Guernsey				
82.	Guinea	6364	25.9	5850	23.8
83.	Guinea-Bissau	1972	70.1	224	8.0
84.	Guyana	16526	84.0	1795	9.1
85.	Haiti	97	3.5		
86.	Honduras	4592	41.0	1619	14.5
87.	Hungary	2069	22.7	121	1.3
88.	Iceland	49	0.5	144	1.4
89.	India	70682	23.8	4124	1.4
90.	Indonesia	91010	53.0	26748	15.6
91.	Iran (Islamic Republic of)	10692	5.8	3627	2.0
92.	Iraq	825	1.9	259	0.6
93.	Ireland	754	10.9	47	0.7
94.	Isle of Man	3	6.1		
95.	Israel	165	7.6	60	2.8
96.	Italy	9297	31.6	1813	6.2
97.	Jamaica	335	31.0	187	17.3
98.	Japan	24958	68.5		

S.No	Country/Territory	Forest		Other wooded land	
		1000 ha	% of land area	1000 ha	% of land area
99.	Jersey	1	5.2		
100.	Jordan	98	1.1	51	0.6
101.	Kazakhstan	3309	1.2	16479	6.1
102.	Kenya	4413	7.8	9365	16.5
103.	Kiribati	12	15.0		
104.	Kuwait	6	0.4		
105.	Kyrgyzstan	637	3.3	704	3.7
106.	Lao People's Democratic Republic	18761	81.3	2846	12.3
107.	Latvia	3356	54.0	112	1.8
108.	Lebanon	137	13.4	106	10.4
109.	Lesotho	49	1.6	96	3.2
110.	Liberia	4179	43.4		
111.	Libya	217	0.1	330	0.2
112.	Liechtenstein	7	43.1	1	3.1
113.	Lithuania	2180	34.8	104	1.7
114.	Luxembourg	87	33.5	1	0.5
115.	Madagascar	12473	21.4	14466	24.9
116.	Malawi	3147	33.4		
117.	Malaysia	22195	67.6		
118.	Maldives	1	3.3		
119.	Mali	4715	3.9	19406	15.9
120.	Marshall Islands	13	70.2		
121.	Martinique	49	45.8	1	1.0
122.	Mauritania	225	0.2	3040	2.9
123.	Mauritius	39	19.2	12	5.9
124.	Mayotte	6	15.5	1	2.8
125.	Mexico	66040	34.0	19715	10.1
126.	Micronesia (Federated States of)	64	91.9		
127.	Mongolia	12553	8.1	5125	3.3

S.No	Country/Territory	Forest		Other wooded land	
		1000 ha	% of land area	1000 ha	% of land area
128.	Montenegro	827	61.5	137	10.2
129.	Montserrat	3	25.0	2	16.0
130.	Morocco	5632	12.6	580	1.3
131.	Mozambique	37940	48.2	14421	18.3
132.	Myanmar	29041	44.2	15080	22.9
133.	Namibia	6919	8.4	8107	9.8
134.	Nepal	3636	25.4	1897	13.2
135.	Netherlands	376	11.1		
136.	Netherlands Antilles	1	1.5	33	40.9
137.	New Caledonia	839	45.9	371	20.3
138.	New Zealand	10152	38.6	1060	4.0
139.	Nicaragua	3114	25.9	2219	18.4
140.	Niger	1142	0.9	3140	2.5
141.	Nigeria	6993	7.7	2681	2.9
142.	Niue	18	69.6		
143.	Northern Mariana Islands	29	64.1		
144.	Norway	12112	39.8	2012	6.6
145.	Occupied Palestinian Territory	9	1.5		
146.	Oman			1303	4.2
147.	Pakistan	1472	1.9	1521	2.0
148.	Palau	40	87.6		
149.	Panama	4617	62.1	532	7.2
150.	Papua New Guinea	33559	72.5	4087	8.8
151.	Paraguay	15323	38.6		
152.	Peru	73973	57.8	9403	7.3
153.	Philippines	8040	27.0	7036	23.6
154.	Pitcairn	4	74.5	1	10.6
155.	Poland	9435	30.8		
156.	Portugal	3182	35.3	1725	19.1

S.No	Country/Territory	Forest		Other wooded land	
		1000 ha	% of land area	1000 ha	% of land area
157.	Puerto Rico	496	57.9		
158.	Republic of Korea	6184	63.7		
159.	Republic of Moldova	409	12.4	74	2.3
160.	Reunion	88	35.2	51	20.4
161.	Romania	6861	29.8	90	0.4
162.	Russian Federation	814931	49.8	74925	4.6
163.	Rwanda	480	19.5	61	2.5
164.	Saint Helena, Ascension and Tristan da Cunha	2	5.1		
165.	Saint Kitts and Nevis	11	42.3	2	8.5
166.	Saint Lucia	20	33.3	9	14.1
167.	Saint Martin (French part)	1	21.2	2	36.1
168.	Saint Pierre and Miquelon	3	12.2		
169.	Saint Vincent and the Grenadines	27	69.2		
170.	Saint-Barthelemy			1	23.8
171.	Samoa	171	60.4	22	7.8
172.	Sao Tome and Principe	54	55.8		
173.	Saudi Arabia	977	0.5	1117	0.5
174.	Senegal	8273	43.0	5250	27.3
175.	Serbia	2720	31.1	508	5.8
176.	Seychelles	41	88.4		
177.	Sierra Leone	3044	42.5	200	2.8
178.	Singapore	16	23.4	4	6.3
179.	Slovakia	1940	40.3		
180.	Slovenia	1248	62.0	23	1.1
181.	Solomon Islands	2185	78.1	129	4.6
182.	Somalia	6363	10.1		
183.	South Africa	9241	7.6	24558	20.2
184.	South Sudan	7157	11.3	32582	51.6

S.No	Country/Territory	Forest		Other wooded land	
		1000 ha	% of land area	1000 ha	% of land area
185.	Spain	18418	36.9	9209	18.5
186.	Sri Lanka	2070	33.0	52	0.8
187.	Sudan	19210	10.3	20677	11.1
188.	Suriname	15332	95.4		
189.	Swaziland	586	34.1	496	28.8
190.	Sweden	28073	68.4	2432	5.9
191.	Switzerland	1254	31.4	70	1.8
192.	Syrian Arab Republic	491	2.7	35	0.2
193.	Tajikistan	412	3.0	142	1.0
194.	Thailand	16399	32.1		
195.	The former Yugoslav Republic of Macedonia	998	39.6	143	5.7
196.	Timor-Leste	686	46.1		
197.	Togo	188	3.5	1246	22.9
198.	Tonga	9	12.5	1	1.4
199.	Trinidad and Tobago	234	45.7	129	25.1
200.	Tunisia	1041	6.7	293	1.9
201.	Turkey	11715	15.2	10130	13.2
202.	Turkmenistan	4127	8.8		
203.	Turks and Caicos Islands	34	80.0		
204.	Tuvalu	1	33.3		
205.	Uganda	2077	10.4	2879	14.4
206.	Ukraine	9657	16.7		
207.	United Arab Emirates	323	3.9	5	0.1
208.	United Kingdom	3144	13.0	20	0.1
209.	United Republic of Tanzania	46060	52.0	7984	9.0
210.	United States of America	310095	33.8	21279	2.3
211.	United States Virgin Islands	18	50.3		
212.	Uruguay	1845	10.5		

S.No	Country/Territory	Forest		Other wooded land	
		1000 ha	% of land area	1000 ha	% of land area
213.	Uzbekistan	3220	7.3	115	0.3
214.	Vanuatu	440	36.1	476	39.0
215.	Venezuela (Bolivarian Republic of)	46683	52.9	7847	8.9
216.	Vietnam	14773	47.6	456	1.5
217.	Wallis and Futuna Islands	6	41.7	4	26.3
218.	Western Sahara	707	2.7		
219.	Yemen	549	1.0	1406	2.7
220.	Zambia	48635	65.4	6108	8.2
221.	Zimbabwe	14062	36.4		

(FAO, 2015b)

CLASSIFICATION OF WORLD FOREST

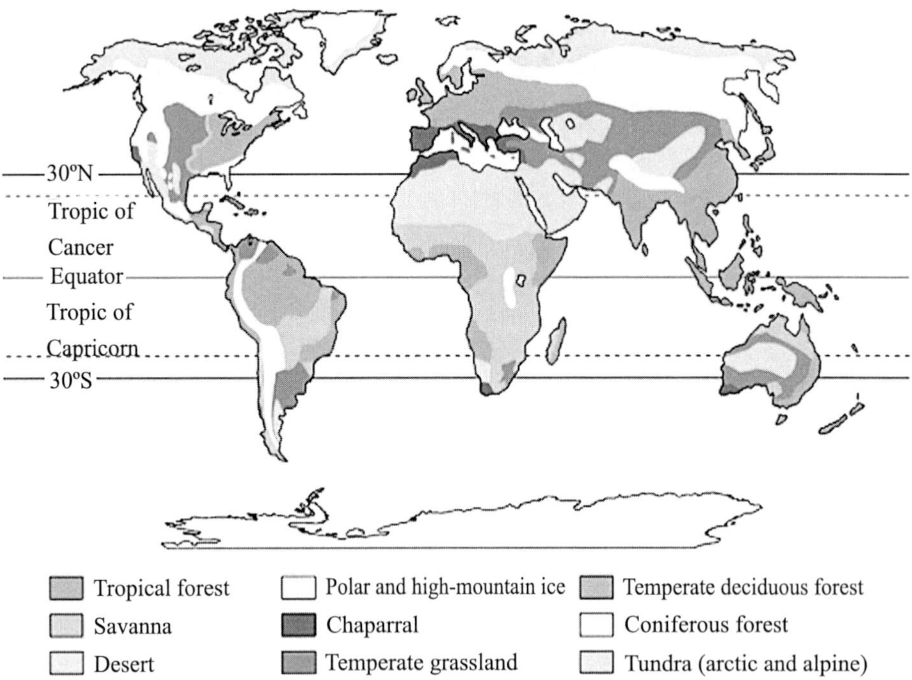

The forest distributed across the world are classified into 9 major types as indicated below.

1. Tropical forest
2. Savannah
3. Desert
4. Polar and high mountain ice
5. Chaparral
6. Temperate grassland
7. Temperate deciduous forest
8. Coniferous forest
9. Tundra (Arcic and Alpine)

Tropical Forest

- These forests are occupied in the area between tropic of cancer and tropic of capricorn.
- These tropical forests include the rainforest distributed in latin America, Africa, Indo Malayan region and also in Australia.
- These tropical forests cover almost 10% of the land above mean sea level (MSL).
- Among the distributed tropical forest, the amazon forest contribute 1/3 of the tropical forest.

Savannah

- Savannahs are the tropical grasslands in which trees are distributed sporadically throughout the tropical region.
- These tropical savannahs include both woodlands and grasslands because the ecosystem processes and management issues are almost similar across both landscapes.
- The tropical savannahs are distributed predominantly in Africa, Australia and South East Asia.
- The tropical savannahs cover little less than a 3rd of the world land surface.
- In Australia, it covers 1/3 of the land area and in South America it covers nearly 45% of the geographic area, in India and South East Asia it covers almost 10% of the total area.

Desert

- The area which receives less than 10 inches of rainfall coupled with extreme dryness are considered to be deserts.

- The presence of plants such as cacti, sage brush, and prosopis are the characteristic feature of desert area.
- The animals like mammals, lizards, snakes, insects and even birds are also forms the part of desert ecosystem.
- The extreme weather conditions, high rate of evaporation, poor productivity etc are the characteristic features of the desert area.
- The deserts are predominantly distributed in Africa, Australia, North America and Asia.

Polar and High Mountain Ice

- This type is the coldest place on the earth and the average temperature is always below 0°C
- The most important feature of this type is the presence of permanent ice and as a result the vegetation growth is almost physically absent.
- This region is categorized by the presence of some of the microscopic algae.
- Animals like polar bear and penguins are commonly present in this region.
- Herbivores do not exist in this type due to the absence of vegetation.#

Chaparral

- This is one of the unique land use system and receives the annual rainfall of 64-76 cm and falls predominantly in winter.
- The summer season is very dry which results in complete dormancy in most of the vegetation.
- The chaparrals are found in California (North America), around Mediterranean Sea and along the southern coast of Australia.
- The vegetation is characterized by the presence of oak both deciduous and evergreen type.
- Presence of scrub oaks and shrubs like Manzanita and the California lilac are evidenced in the form of dense evergreen thickets.

Temperate Grassland

- These grasslands are known as Prairie and receives an annual rainfall of 20 inches per annum.
- Fire is one of the major causative factors that tips the balance from forest to grasslands.
- Wide ranges of animals are present and among them bisons and antilops were dominant.

- The presences of vertibrates are also witnessed.
- This temperate grassland also supports cultivation of corn, wheat and other grains.
- Domestication of cattle and sheep is common in this grasslands.

Temperate Deciduous Forest

- This forest type occupies the eastern half of United States and a large portion of Europe.
- This forest type harbours hardwoods liken beaches, maples, oaks and hickory.
- One of the unique characteristics of this forest type is the large stands are dominated by a single species.
- The major animals in this habitat include deers, raccoons and salamanders.
- This temperate deciduous forest is quite productive in both natural and agriculture ecosystem.

Coniferous Forest

- This forest type occurs between the tundra in the north and deciduous forest in the south.
- The northern boreal forest occurs between 50° to 60° N latitude.
- The temperate coniferous forest occurs in lower latitude of North America, Europe and Asia in the higher elevation of mountain.
- This forest type harbours predominantly conifers like pines, spruce and fir.
- The precipitation varies from location to location and ranged between 300 and 2000 mm.

Tundra (Artic and Alpine)

- It is one of the unique vegetations where in the trees of taiga becomes stunted by the harshness of subarctic climate.
- The climate is very cold in winter.
- The landscape is dominated by the presence of wide range of lichens, grasses and some fast growing animals.
- The tundra vegetation harbours the swarms of migrating birds especially water fowl.

■■■■

CHAPTER - 4

FOREST RESOURCES OF IMPORTANT COUNTRIES

The information on global forest resources particularly the extent of forest cover across the major countries is very important. The information on global forestry resources needs to be updated in order to assess the changes that happen in the forest cover. The first comprehensive global forest resource assessment was made during the year 1923 but this assessment did not include all the countries. Hence FAO has conducted global forest assessment since 1948 and the recent assessment was made in the year 2015. This chapter elaborates the forest resources of the following countries.

- Western Europe
- North America
- Central America
- Central Africa
- Australia
- Russia
- China
- Japan

I. FOREST RESOURCES OF WESTERN EUROPE

The Western Europe is endowed with fourteen different forest types ranging from boreal forest to mountain forest.

FOREST TYPES AND SPECIES COMPOSITION OF WESTERN EUROPE

S.No	Forest Type	Species Composition
I.	Boreal forest	• Picea - *Picea abies* • Pinus - *Pinus sylvestris* • Birches - Betula spp. • Aspen - *Populus tremula* • Rowan - *Sorbus aucuparia* • Willows - Salix spp.

S.No	Forest Type	Species Composition
II.	Hemiboreal forest and Nemoral coniferous and mixed broad leaved - coniferous forest	• Oak - *Quercus robur* • Ash - *Fraxinus excelsior* • Ulm - *Ulmus glabra* • *Tilia cordata*
III.	Alpine coniferous forest	• Larch - *Larix deciduas* • *Pinus cembra* • *P. nigra and* • *P. mugo*
IV.	Acidophilous oak and oak-birch forest	• *Acidophilous oaks* • *Q. robur* • *Q. petraea* • Birch - *Betula pendula*
V.	Mesophytic deciduous forest	• *Carpinus betulus* • Oak - *Quercus petraea* • Oak - *Quercus robur* • Ash - *Fraxinus,* • *Acer* spp. • *Tilia cordata*
VI.	Beech forest	• Beech - *Fagus sylvatica* • *Fagus orientalis* • *Betula pendula*
VII.	Mountainous beech forest	• Fagus • *Betula pendula*
VIII.	Thermophilous deciduous forest	• Quercus • Acer • Ostry • Fraxinus • Carpinus
IX.	Broadleaved evergreen forest	• Oak • Birch • Beech
X.	Coniferous forests	• Pinus • *P. halepensis* • *P. canariensis*

S.No	Forest Type	Species Composition
		• Abies • Juniper
XI.	Mire and swamp forest	• *Picea abies* • *Pinus sylvestris* • Alnus • Betula • Quercus • Populus
XII.	Floodplain forest	• Alnus • Betula • Populus • Salix • Fraxinus • Ulmus. • *Fraxinus angustifolia* • *Nerium Oleander* • *Platanus orientalis* • Tamarix
XIII.	Non riverine alder, birch, or aspen forest	• Alnus • Betula • Populus
XIV.	Plantations and self sown exotic forest	• **Forest types** • Plantations of site-native species • Plantations of not-site-native species and self-sown exotic forest • **Species Composition** • The forest predominantly consists of self-sown non-native, often invasive, tree species.

Forest Resources North America

- North America covers an area of about 24,709,000 square kilometers (9,540,000 square miles), about 16.5% of the earth's land area and about 4.8% of its total surface.
- The United States and Canada together have about 15.5 percent of the world's total forest cover.

	Canada	US	Mexico
Total Forest Land (1000 ha)	401530	303074	55300
Total Harvest (Mill.m^3)	192.1	452.1	41.3

Canada

Forests are primarily publicly owned.

- They are mostly natural forests under extensive management.
- Access to fiber is through licensing arrangements with provincial governments.

US

57% of forestland is publicly owned.

- National forestland is not important for industrial forestry.
- Commercial forestry is primarily practiced on private forestland mostly owned by non-industrial land owners, who supply 60% of the total harvest.
- Industrial timber plantations cover 18 mill.ha.

Mexico

80% of forestland is common property land owned by edijos or comunidades forestales.

- Mexican forests contain most pine and oak species of the world.

Fourteen different forest types found in North America are furnished below

1. Tundra
2. Boreal Forest
3. Rocky mountain evergreen forest
4. Pacific coast evergreen forest
5. Northern mixed forest
6. Eastern deciduous forest
7. Coastal plain mixed evergreen forest
8. Mexican montane forest
9. Central american rain forest
10. Great plains grasslands
11. Tropical savanna

12. Cold Desert
13. Hot Desert
14. Mediterranean Shrub

Tundra

- Tundra means rolling treeless plains. The normal weather is frigid and dry winters with cold and moist summers. In North America, arctic tundra is found in northern **Alaska, Canada, and Greenland**.
- The two most common trees are arctic willow and diamond-shaped willow. Deep-rooted trees cannot survive due to the permafrost.
- Alternate names for this biome are alpine tundra, wet tundra, and dry tundra.

Boreal Forest

- This forest is located in the very northern portion of North America including most of Canada and a top high-altitude mountains in the United States. The normal weather is frigid, long, and dry winters with short, cool, and moist summers.
- Fir, spruce, larch, aspen, and jack pine and dominant species.
- The boreal forest separates the tundra from the temperate forest.
- The alternative names for the boreal forest biome are subalpine and taiga.

Rocky Mountain Evergreen Forest

- Montane is a term for a forest on mid-level elevations on mountains. The normal weather is cold and moist winters with mild and moist summers.
- Douglas fir forests with western white pine, western larch, grand fir, and western ponderosa pine are the major species.
- Alternate names are Rocky Mountain steppe and montane forest.

Pacific Coast Evergreen Forest

- It is interesting to note that this is one of the largest temperate rain forests.
- The Pacific temperate rain forests lie west of the Pacific mountain ranges, from southern Alaska to northern California.
- The normal weather is mild and very moist winters with mild and moist summers.

- Trees include Sitka spruce, Douglas fir, redwood, western red cedar, alder and bigleaf maple.
- The alternate name is temperate rain forest.

Northern Mixed Forest

- The normal weather is cold and moist winters with mild and moist summers. The species distrbuted in this forest are Beech, Maple, Eastern hemlock, Yellow birch, White pine, and Northern white cedar.
- The alternate names are northern hardwood-hemlock and transitional mixed forest.
- The majority of the trees drop their leaves at the end of the typical growing season in a deciduous forest.

Eastern Deciduous Forest

- This biome is found east of the Mississippi River. The normal weather is cool/cold and moist winters with warm and moist summers.
- Beech, Maple, Yellow Poplar, Oak - Hickory, Mixed pine-hardwood are the major species food in this forest.
- The alternate name is transitional mixed forest.

Coastal Plain Mixed Evergreen Forest

- Atlantic and Gulf coastal plains down to the sea.
- There often are sluggish streams, marshes, and swamps. The normal weather is cool/mild and moist winters with hot and moist summers.
- The trees include one beech, maple, yellow poplar, oak, hickory, and mixed pine-hardwood.
- The alternate name is southeastern mixed evergreen forest.

Mexican Montane Forest

- These forests are found in the mountains of Mexico.
- The alternate names are tropical montane forest and cloud forest.
- The normal weather is mild and dry winters with mild and moist summers.
- There are wide variety of species and many of them one are unique.

Central American Rain Forest

- The alternate names are tropical rain forest and selva.

- The normal weather is warm and very moist winters with hot and very moist summers.
- There are a vast array of species of trees.

Great Plains Grasslands

- Oak, maple, hackberry, dogwood, cottonwood, and cedar can be found in Great Plains grasslands, especially in river systems.
- The normal weather is cool/frigid and dry winters with hot and moist summers. The alternate names are prairie and steppe.

Tropical Savannah

- The normal weather is warm and very dry winters with hot and moist summers. Tropical savannah is dominated by grasses.
- Alternate names are West Indian savannah, tropical thorn scrub, tropical dry forests, and Florida Everglades.

Cold Desert

- The alternate names are Great Basin desert and high plain.
- The normal weather is cold and dry winters with warm and very dry summers. Plants are widely scattered and sagebrush often predominates.
- In semi-arid places, plants include creosote bush, bur sage, white thorn, cat claw, and mesquite

Hot Desert

- These areas include the Sonoran, Mojave and Chihuahua deserts.
- The normal weather is mild winters and very dry winters with hot and very dry summers.
- Plants are mainly short shrubs and short woody trees. Plants include yuccas, ocotillo, turpentine bush, prickly pears, false mesquite, agaves, and brittlebush.

Mediterranean Shrub

- The alternate name is Californian chaparral.
- The normal weather is mild and moist winters with warm and very dry summers.
- Trees include oak, pine, and mahogany. North facing slopes get more moisture and may have manzanita, toyon, scrub oak, and poison oak.

- South facing slopes are drier and may have chamise, black sage, and yucca.

Industrial Forest Production and Trade in North America

- Although forest products production totalled USD 240 billion in the US and USD 42 billion in Canada, forestry has a relatively more important role in the economy of Canada.
    - Forest products trade between Canada and the US is the largest in the world. 90% of Canada's lumber production (volume) goes to the US markets.
    - Access to US markets has major implications for forest management in Canada.
- Mexico's total timber production is only about 2% of the US's and 5% of Canada's production volumes. Almost 80% of timber harvest is for fuelwood. Mexico and the US are net-importers of timber.
- Non-timber forest products (NTFPs) have an important role in North American culture and commerce, although they are usually regarded as by-products of forest management.
- Varying degrees of forest values and practices are evolving to encompass economic, environmental, social and cultural considerations.
    - Forest management has developed from a focus on harvest and regeneration issues (fiber provision), through multiple use management, towards the current paradigms of sustainable forest management (SFM) and/or ecosystem management.
    - SFM is becoming an explicit forest policy goal across North America. Ecosystem management is still a highly debated normative concept

Forest Resources of Central America

Central America is bordered by Mexico to the north, Colombia to the southeast, the Caribbean Sea to the east, and the Pacific Ocean to the west. Central America consists of seven countries *viz.,* Belize, Costa Rica, El Salvador, Guatemala, Honduras, Nicaragua and Panama

Land area: 507,966 km^2 (196,127 sq mi)

Biodiversity in Central America

- Central America is part of the Mesoamerican biodiversity hotspot, boasting 7% of the world's biodiversity.

- The Pacific Flyway is a major north-south flyway for migratory bird in the Americas, extending from Alaska to Tierra del Fuego.
- Due to the funnel-like shape of its land mass, migratory birds can be seen in very high concentrations in Central America, especially in the spring and autumn.
- As a bridge between North America and South America, Central America has many species from the Neoarctic and the Neotropic eco zones.
- However the southern countries (Costa Rica and Panama) of the region have more biodiversity than the northern countries (Guatemala and Belize), meanwhile the central countries (Honduras, Nicaragua and El Salvador) have the least biodiversity.

FORESTS OF CENTRAL AMERICA

The Central American Forest consists of eight different forest types varying from tropical forests to mangrooves. The forest hypel and species composition are furnished below:

Forest Types and Species Composition of Central America

S.No	Forest Type	Species Composition
I.	Tropical and subtropical moist broadleaf forests	• Maple • Oak
II.	Tropical and subtropical dry broadleaf forests	• Tropical cedar - *Cedrela* spp. • Pine, • Juniper, • Oak
III.	Tropical and subtropical coniferous forests	• Pine - Pinus spp. • Douglas fir - Pseudotsuga spp. • Cedar - Juniperus spp. • Cupressus spp. • Thuja spp. • Fir - Abies spp. • Redwood - Sequoia spp.
IV.	Tropical and subtropical grasslands, savannas, and shrublands	• Shallot species: *A. oschaninii* • Valerian species: *C. ruber* • Baobab species: *Adansonia* • Acacia species: about 1,300 species

V.	Flooded grasslands and savannas	• Northern white cedar - *Thuja occidentalis*) • *Betula papyrifera* - Paper birch • *Cornus florida* - Flowering dogwood • *Larix laricina* - Tamarack • *Picea mariana* - Black spruce • *Picea glauca* - White spruce • *Pinus strobes* - White pine • *Tsuga Canadensis* - Hemlock • *Ulmus Americana* - American elm
VI.	Montane grasslands and shrublands	• Steraceae (*Psiadia* spp., *Helichrysumspp*., Stoebe spp., *Stenocline* spp.) • Ericaceae (*Erica* spp., *Agauria* spp., *Vaccinum* spp.) • Podocarpaceae (*Podocarpus* spp.) • Rhamnaceae (*Phylica* spp.) • Rubiaceae plant families.
VII.	Deserts and xeric shrublands	• Crucifixion Thorn • Cottonwood • Creosote Bush • Desert Lavender • Desert Willow • Juniper • Ocotillo • Mesquite Tree • Tumbleweed • Whitethorn Acacia
VIII.	Mangrove	• Red mangrove (*Rhizopora mangle, R. harrisonii*) • Black mangrove (*Avicennia germinans*), • White mangrove (*Laguncularia racemosa*), • Buttonwood (*Conocarpus erectus*) and • Leather fern (*Acrostichum* sp)

FOREST TYPES OF CENTRAL AMERICA - FAO CLASSIFICATION

Pine-Oak Forests

- The Central American pine-oak forests eco region, in the tropical and subtropical coniferous forests biome, is found in Central America and southern Mexico.

- The Central American pine-oak forests occupy an area of 111,400 square kilometers (43,000 sq mi), extending along the mountainous spine of Central America, extending from the Sierra Madre de Chiapas in Mexico's Chiapas state through the highlands of Guatemala, El Salvador, and Honduras to Central Nicaragua.
- The pine-oak forests lie between 600–1,800 metres (2,000–5,900 ft) elevation, and are surrounded at lower elevations by tropical moist forests and tropical dry forests.
- Higher elevations above 1,800 metres (5,900 ft) are usually covered with Central American montane forests.
- The Central American pine-oak forests are composed of many species characteristic of temperate North America including oak, pine, fir, and cypress.

Laurel Forest

- Laurel forest is the most common type of Central American temperate evergreen cloud forest, found in almost all Central American countries, normally more than 1,000 meters (3,300 ft) above sea level.
- Tree species include evergreen oaks, members of the laurel family, and species of *Weinmannia*, *Drimys*, and *Magnolia*

Cloud Forest

- The cloud forest of Sierra de las Minas, Guatemala, is the largest in Central America.
- In some areas of southeastern Honduras there are cloud forests, the largest located near the border with Nicaragua.
- In Nicaragua, cloud forests are situated near the border with Honduras, but many were cleared to grow coffee.
- There are still some temperate evergreen hills in the north. The only cloud forest in the Pacific coastal zone of Central America is on the Mombacho volcano in Nicaragua.

Central American Montane Forests

- The Central American montane forests are an ecoregion of the tropical and subtropical moist broadleaf forests biome.
- These forests are of the moist deciduous and the semi-evergreen seasonal subtype of tropical and subtropical moist broadleaf forests and receive high overall rainfall with a warm summer wet season and a cooler winter dry season.

- Central American montane forests consist of forest patches located at altitudes ranging from 1,800–4,000 metres (5,900–13,100 ft), on the summits and slopes of the highest mountains in Central America ranging from Southern Mexico, through Guatemala, El Salvador, and Honduras, to northern Nicaragua.
- The entire ecoregion covers an area of 13,200 square kilometers (5,100 sq mi) and has a temperate climate with relatively high precipitation levels.

Forest Resources of Central Africa

- Central Africa groups 7 countries which are covered partly or totally by the large stretch of humid forest of the central western part of Africa.
- The countries are:
    - Angola
    - Cameroon
    - Central African Republic
    - Congo
    - Gabon
    - Equatorial Guinea
    - Zaire
- The Central African Republic (CAR) is located in the geographical heart of Africa and on the northwestern limit of the Congo Basin.
- Covering a land surface area of 623,000 square kilometers, it is approximately one and a half times the size of France.
- Most of the country's surface area is covered by wooded savannah.
- However the southern part of the country consists of a band of several million hectares of rain forest much of which is now being exploited by commercial logging.
- Another natural resource is diamonds which are found throughout the country and contribute more than 70% to the total national economy.

Forest Types in Central Africa

Salinet Features

- Tropical African forest is 18 per cent of the world total and covers over 3.6 million square kilometers of land in West, East and Central Africa.
- This total area can be subdivided to 2.69 million square kilometers (74%) in Central Africa.

- The main vegetation features of Central African rain forest divides the area into the following categories:
- This type of forest shows no substantial seasonal behavior.
- At the border of the central basin is the mesophilous semi-deciduous forest that is mixed with deciduous and evergreen trees in the upper-stratum, unusual age distribution, continuous shrub stratum at the lower canopy, and a more marked seasonality.

Forest Types

Secondary Forests

- Beyond the forest reserves, a great deal of the remaining part of Central African rain forest on well drained soils are old secondary forests.
- There also exist younger secondary forests dominated by parasol trees, ***Musanga ceropiodes***, the most abundant and characteristic secondary forest in Africa.
- Such trees are found in upper layers of secondary re-growth along the old road networks in Zaïre .
- The dispersal of secondary forests are important in regional study as they show different floristic and faunistic characteristics than primary forests, and represent centers of human activity and history of land-use changes.

Non-Forest

- The part comprises degraded lands, irregular agriculture, plantations, deforested lands and fragmented forests.
- Plantations have a geometrical lay-out with uniform canopy-cover and follow a different vegetation cycle as their adjacent features.
- The areas are located near the transportation networks and are subjugated by economic tree crops such as cacao, rubber, oil palms and Kola.

Swamp and Flooded Forests

- Swamp forest, inundated forests in flood plains, and riparian forests are found here.
- Swamp forests are found widely in the Zaïre basin and throughout the Congo basin where conditions are appropriate.
- In most areas, swamp forests is like in appearance to rain forest and the tallest trees attain a height of 45 m.

- The main canopy is often irregular and open, sometimes resembling the secondary forests caused by disturbance The forest has a variety in endemic flora but it is inadequate in species.
- Recently, large areas of swamp forests have been cleared for rice farming.
- Swamp forests in Zaïre, Congo and other lowland forests have seasonal variations that to the level of forest inundation.

Central African Rain Forest

- Central Africa's rainforests are second only to those of the Amazon basin in terms of size.
- The rainforest supports an estimated 8,000 plant species but among the best-known are the oil palm and the mahogany tree.

CENTRAL AFRICAN RAIN FOREST TYPES

Congolese Forests

- The **Congolese forests** are a broad belt of lowland tropical moist broadleaf forest eco region of the Tropical and subtropical moist broadleaf forests Biome, which extends across the basin of the Congo River and its tributaries in Central Africa.
- The Congolese forests cover southeastern Cameroon, eastern Gabon, the northern and central Republic of the Congo, the northern and central Democratic Republic of the Congo and portions of southern and southwestern Central African Republic

Western Congolese Swamp Forests

- The ecoregion contains areas of permanently flooded swamp forest, seasonally flooded swamp forest, and flooded grassland.
- The permanently flooded swamp forests are home to extensive stands of *Raphia* palm. Trees in the seasonally flooded forests include species of *Garcinia* and *Manilkara.*

The Western Congolose swamp forest are further classified in to 14 different types.

Central Congolian Lowland Forest

- *Gilbertiodendron dewevrei*
- *Julbernadia seretii*
- *Brachystegia laurentii*

- Hackberry (*Celtis* spp)
- Samfona (*Chrysophyllum perpulchrum*) and
- Bark cloth tree (*Antiaris welwitschii*)

Cameroon Highland Forest

- Also known as Mossy Forest

Species

- *Nuxia congesta*
- *Podocarpus latifolius*
- *Prunus africana*
- *Rapanea melanophloeos*
- *Syzygium guineense bamendae*

Central African Mangrove

- Red mangrove (*Rhizophora mangle*)
- *Rhizophora racemosa*
- *Rhizophora harrisonii*
- Black mangrove (*Avicennia germinans*)
- White mangrove (*Laguncularia racemosa*)
- Palm - *Nypa fruticans*

Congolian Coastal Forest

- *Ocotea pretiosa* and *O. catharinense* (Lauraceae)
- *Campomanesia xanthocarpa* (Myrtaceae)
- *Mimosa scabrella* and
- *Parapiptadenia rigida* (Leguminosae)
- Brazilian araucaria (*Araucaria angustifolia*)

Congolian Riverine Forest

- *Acacia elatior* (River Acacia)
- *Acacia galpinii* (Monkey thorn)
- *Acacia gerrardii* (Grey-haired acacia)
- *Albizia glaberrima* (White Nongo)
- *Albizia zygia* (West African albizia)

- *Allophylus africanus*
- *Cadaba farinosa* (Herd's boy fruit)
- *Cordyla africana* (Wild mango)
- *Dovyalis abyssinica* (African gooseberry)
- *Kigelia africana* (Sausage tree)
- *Newtonia buchananii* (East African newtonia)
- *Piliostigma thonningii* (Camel's foot)

Congolian Swamp Forest

- The forest is a mixture of habitats including wetlands and swamps, with drier forest and savanna slightly higher and flooded seasonally by the Congo and its tributaries.
- Raphia palm
- Garcinia
- Manilkara

Cross River Forest

- *Musanga cecropioides* (African corkwood tree or umbrella tree)
- *Irvingia gabonensis*
- *Berlinia confusa*
- *Coula edulis*
- *Hannoa klaineana*
- *Klainedoxa gabonensis*
- *African mahoganey*
- Red ironwood - *Lophira alata*

East Sudanian Savanna

- Senegal date palms *(Phoenix reclinata)*
- Wild custard apple trees *(Annona senegalensis)*
- African breadfruit *(Treculia africana)*
- *Afzelia africana*
- *Chrysophyllum perpulchrum*
- *Cola gigantea*
- *Morus mesozygia*
- *Khaya grandifoliola*

- *Burkea africana*
- *Stereospernum kunthianum*
- *Strychnos* spp.
- V*itex* spp.

Mount Cameroon Forest

Chief plant communities are

- Montane forests
- Montane grasslands
- Heathlands

Species

- *African Zebrawood*
- *Microberlinia bisulcata*
- *Cola praeacuta* and
- *Belonophora ongensis.*

NE congolian FSM

- *Julbernardia*
- *Cynometra species*
- *Gilbertiodendron dewevrei*
- *Entandrophragma*
- *Khaya species*
- *Albizia*
- *Musanga cecropiodes*

NE Congolian Lowland Forest

- *Khaya nyasica*
- *Mussaenda erythrophylla*
- *Ficus zarzalensis* – Stranger Fig
- *Pericopsis elata*
- Caesalpiniaceae trees.
- Okoumé trees (Burseraceae)

NW Congolian Lowland Forest

- *Entandrophragma congoense*
- *Pentaclethera eetveldeana*
- *Pericopsis elata*
- *Gilbertiodendron dewevrei*
- *Drypetes calvescens*
- *D. capillipes*
- *D. chevalieri*

South Congolian FSM

- *Khaya nyasica*
- *Mussaenda erythrophylla*
- *Pentaclethera eetveldeana*
- *Chrysophyllum perpulchrum*
- *Khaya grandifoliola*
- *Burkea africana*
- *Newtonia buchananii*

West Congolian FSM (Forest Savannah Mosaic)

- *Musanga ceropiodes*
- *Manilkara jaimiqui*
- *Parapiptadenia rigida*
- *Musanga ceropiodes*
- *Khaya nyasica*
- *Khaya grandifoliola*
- *Newtonia buchananii*

V. FOREST RSOURCES OF AUSTRALIA

Salient Features

- Australia has 125 million hectares of forest, which is 16 per cent of Australia's land area.
- This is about 3 per cent of the world's forest area.
- Seventh-largest reported forest area worldwide.

Australia – Forest Area

- Queensland has the largest area of Australia's forest (51.0 million hectares — 41% of Australia's forest)
- New South Wales (22.7 million hectares—18%),
- Western Australia (19.2 million hectares—15%)
- Northern Territory (15.2 million hectares—12%)
- South Australia (17.8 million hectares—14%)

FOREST TYPES – BASED ON NATIONAL FOREST INVENTORY

- Australia's forests are classified in the National Forest Inventory into three categories:
  1. Native forest
  2. Industrial plantations
  3. Other forest

Native Forest

Australia's native forests are grouped into eight types (More than 80 per cent of Australia's native forest area is dominated by Eucalypt forest and Acacia forest):

(i) Acacia forest

(ii) Callitris forest

(iii) Casuarina forest

(iv) Eucalypt forest

(v) Mangrove forest

(vi) Melaleuca forest

(vii) Rain forest

(vii) Other native forest.

Acacia forest

– *Acacia acinacea* - Gold Dust Wattle

– *Acacia adunca* - Wallangarra Wattle

– *Acacia baileyana* - Cootamundra Wattle

– *Acacia boormanii* - Snowy River Wattle

– *Acacia dimidiata* - Swamp Wattle

– *Acacia extensa* - Wiry Wattle

– *Acacia falciformis* - Hickory Wattle

– *Acacia flexifolia* - Bent-leaf Wattle

– *Acacia glaucoptera* - Flat Wattle

– *Acacia howittii* - Sticky Wattle

– *Acacia montana* - Mallee Wattle

– *Acacia beckleri* - Barrier Range Wattle
– *Acacia buxifolia* - Box-leaf Wattle
– *Acacia conferta* - Crowded-leaf Wattle
– *Acacia cultriformis* - Knife-leaf Wattle
– *Acacia dealbata* - Silver Wattle
– *Acacia continua* - Thorn Wattle
– *Acacia spectabilis* - Mudgee Wattle
– *Acacia pycnantha* - Golden Wattle
– *Acacia pravissima* - Ovens wattle
– *Acacia siculiformis* - Dagger Wattle
– *Acacia verniciflua* - Varnish Wattle

Callitris forest

– *Callitris baileyi*
– *Callitris canescens*
– *Callitris drummondii*
– *Callitris macleayana*
– *Callitris neocaledonica*
– *Callitris oblonga*
– *Callitris pancheri*
– *Callitris sulcata*
– *Callitris verrucosa*

Casuarina forest

- *Casuarina cristata*
- *Casuarina cunninghamiana* – River Sheoak
- *Casuarina equisetifolia* – Australian Pine
- *Casuarina glauca* - Gray Sheoak
- *Casuarina obesa*
- *Casuarina pauper*

Eucalypt forest

- The Eucalypt forest type is the largest component of both woodland forests (64.2 million hectares) and open forests (27.3 million hectares).
- The Rainforest forest type is the largest component of closed forests (2.6 million hectares).
- River red gum (*E. camaldulensis*) is the most widely distributed eucalypt, and is found in all Australian mainland states.
- The forests of south-eastern Australia contain a wide range of dominant eucalypt species, including major commercial timber species such as
    - Mountain ash (*E. regnans*)

- Messmate stringybark (*E. obliqua*)
- Alpine ash (*E. delegatensis*)
- Silvertop ash (*E. sieberi*)
- Blackbutt (*E. pilularis*)
- Spotted gum (*C. maculata*)

- Eucalypt forest in south-western Australia is dominated by
  - Jarrah (*E. marginata*) and
  - Karri (*E. diversicolor*)
- Typical eucalypts of northern Australia include
  - Darwin woollybutt (*E. miniata*) and
  - Darwin stringybark (*E. tetrodonta*)

Mangrove forest

- Red mangrove
  – *Rhizopora mangle*
  – *R. racemosa*
  – *R. harrisonii*
- Black mangrove (*Avicennia germinans*)
- White mangrove (*Laguncularia racemosa*)
- Buttonwood (*Conocarpus erectus*)

Melaleuca forest

– *Melaleuca quinquenervia*
– *Melaleuca cajuputi* subsp. *cumingiana*
– *M. aspalathoides*
– *M. concinna*
– *M. cajuputi*
– *M. uinquenervia*
– *M. lineariifolia*
– *M. bracteata*

Rain forest

- Australian red cedar - *Thuja plicata*
- Australian Teak tree - *Pterocarpus angolensis*
- Knife Leaf wattle - *Acacia cognata*
- Banksia tree - *Banksia dentata*
- Moreton Bay Fig - *Ficus macrophylla*
- Willow Bottlebrush - *Casllistemon salignus*

- Lilly Pilly - *Syzygium smithii*
- Paperbark - *Melaleuca quinquenervia*

Other native forest

- Illawarra Flame Tree (*Brachychiton acerifolius*)
- Antarctic/Australian Beech (*Nothofagus moorei*)
- Paperbark (*Melaleuca quinquenervia*)
- Lilly Pilly (*Syzygium smithii*)
- Moreton Bay Fig (*Ficus macrophylla*)
- *Eremophila* spp.
- Willow Bottlebrush (*Casllistemon salignus*)
- Golden Wattle (*Acacia pycnantha*)
- *Eucalyptus* spp.

Industrial plantations

- Industrial plantations are grown on a commercial scale for wood production.
- Australia's industrial plantations are a major source of commercial wood products.
- About half of Australia's industrial plantation are exotic softwood (predominantly *Pinus radiata*), while the other half are hardwood (predominantly eucalypt species).

Other forest

- Other forest comprises non-industrial plantations and planted forest of various types.

FOREST ECOSYSTEMS IN AUSTRALIA

Dry Sclerophyll Forests

- Spike Wattle (*Acacia oxycedrus*)
- Myrtle Wattle (*Acacia myrtifolia*)
- Queensland Silver Wattle (*Acacia podalyriifolia*)
- Wax-lip Orchid (*Glossodia major*)
- Grey Spider Flower (*Grevillea buxifolia*)
- Silky Grevillea (*Grevillea sericea*)

- Sydney Boronia (*Boronia ledifolia*)
- *Crowea saligna*

Wet Sclerophyll Forests

- Tall, dense eucalypt forests with open, almost continuous canopies, with a layer of lower trees filling the space between the trunks of the dominant trees.
- There will be a lush floor of bright green ferns, many of them are tree ferns. In subtropical conditions there may be some palms in the understorey.

— *Eucalyptus delegatensis* (alpine ash)
— *E. pauciflora* (white sally)
— *E. fraxinoides* (white ash)
— *E. sieberi* (silvertop ash)
— *Acacia dealbata* (silver wattle)
— *Bossiaea foliosa*,
— *Coprosma hirtella*,
— *Daviesia latifolia* (broad-leaved bitter pea),
— *Leucopogon lanceolatus*,
— *Lomatia myricoides* (river lomatia),
— *Olearia megalophylla*,
— *Persoonia silvatica*,
— *Platysace lanceolata*,
— *Senecio linearifolius* (fireweed groundsel),
— *Tasmannia* lanceolata (mountain pepperbush),
— *T. xerophila* (alpine pepper).

Wet Tropical Rainforest

- Wet Tropics rainforests are such uniquely located and productive habitats that some 1164 genera of vascular plants have been recorded, including 13 of the 19 angiosperm families, the most primitive families of flowering plants worldwide.

— Acai Palm (*Euterpe precatoria*)
— Carnauba Palm (*Copernicia prunifera*)
— Walking Palm (*Socratea exorrhiza*)
— Tualang (*Koompassia excelsa*)
— Radiata Pine (*Pinus radiata*)
— Coachwood (*Ceratopetalum apetalum*)
— *Lomandra hystrix*
— Foambark (*Jagera pseudorhus*)
— Blue quandong – (*Elaeocarpus angustifolius*)
— Queensland Hollywood – (*Pittosporum rhombifolium*)
— Diamond leaf pittosporum – (*Auranticarpa rhombifolia*)
— Red cedar – (*Juniperus virginiana*)

Subtropical Rainforest

- Figs (*Ficus benjamina*)
- Red Cedars (*Juniperus virginiana*)
- Black Booyong (*Argyrodendron actinophyllum*)
- Blue Gums (*Eucalyptus globulus*)
- Hoop Pines (*Araucaria cunninghamii*) and
- Brush Box (*Lophostemon confertus*)

Warm Temperate Rainforest

- Coachwood (*Ceratopetalum apetalum*)
- Sassafras (*Sassafras randaiense*)
- Antarctic Beech (*Lophozonia moorei*)
- Red Ash (*Fraxinus pennsylvanica*)
- Native Clerodendrum (*Clerodendrum thomsoniae*)
- Brown Kurrajong (*Commersonia bartramia*)
- Native Laurel (*Anopterus glandulosus*)
- Red-fruited Olive Plum (*Elaeodendron australe*)
- Cheese Tree (*Glochidion ferdinandi*)
- Native Hibiscus (*Hibiscus* spp.)
- Native Olive (*Olea paniculata*)
- Brown Beech (*Hypsizygus tessellatus*)

Cool-Temperate Rainforests

- Myrtle (*Nothofagus cunninghamii*)
- Sassafras (*Atherosperma moschatum*)
- Leatherwood (*Eucryphia lucida*)
- Horizontal (*Anadopetalum biglandulosum*)
- celery-top pine (*Phyllocladus aspleniifolius*)
- Huon pine

Cloud Forests

— Lignum (*Lignum vitae*)
— Basswood (*Tilia americana*)
— *Cyathea rebeccae*
— Tea-tree (*Melaleuca alternifolia*)
— Mountain Silkwood (*Flindersia oppositifolia*)

— Lawyer Vine (*Smilax australis*)
— Sarsaparilla (*Smilax ornata*)
— Hotbark (*Zygogynum howeanum*)
— Pepperwood (*Umbellularia californica*)
— Malletwood (*Rhodamnia argentea*)
— Native Holly (*Ilex opaca*)
— Pimply Ash (*Fraxinus* spp)

Monsoon and Dry Rainforests

- Blue Quandong (*Elaeocarpus angustifolius*)
- Emu Apple (*Owenia acidula*)
- Red Ash (*Fraxinus pennsylvanica*)
- Milkwood (*Sideroxylon inerme*)
- Yellow Tulipwood (*Drypetes lasiogyna*)
- Pink Mahogany (*Dysoxylum oppositifolium*)
- Carpentaria Palm (*Carpentaria acuminata*)
- Northern Grevillea (*Grevillea pteridifolia*)
- Lawyer Vine (*Smilax australis*)

FOREST RSOURCES OF RUSSIA

Salient Features

- On the territory of Russia is 1/4 of the world's reserves of wood.
- The total forest area - 885 million hectares, representing 45% of the total area of the country.
- The stock of wood was in the area of 82 billion cubic meters.
- The main share of coniferous tree species comprise:
  - Pine
  - Spruce
  - Larch
  - Cedar

Forest type of Russia

Spruce and Fir Forest

- Spruce, fir and Siberian pine dominate, often with presence of birch, aspen, pine and larch.
- Other broadleaved deciduous species and Korean pine are present along the southern border of European Russia and in the south of the Russian Far East.

Dominant species

— *Picea engelmannii* (Engelmann Spruce)

— *Abies lasiocarpa* (Subalpine Fir)

— *Pinus sibirica* (Siberian Pine)

Other species

— *Pinus resinosa*

— *Pinus koraiensis* (Korean Pine)

Pine Forest

- Scotch pine dominates, usually with presence of spruce, birch and aspen and/or larch at the northeastern edge.

— *Pinus sylvestris* (Scotch pine)

— *Picea glehnii* (Sakhalin spruce or Glehn's spruce)

— *Betula platyphylla* (Siberian silver birch)

— *Populus tremuloides* (Aspen)

— *Larix decidua* (European larch)

Larch Forest

- Larch of different species dominates, often with presence of birch and aspen.
- Other species (pine, spruce, Siberian pine) may be mixed along the southern and western borders of Russia.
- In the mountains of the Russian Far East, larch species often have undergrowth of Siberian pine.
- *Larix sibirica* (Russian larch)
- *Larix laricina*
- *Larix occidentalis*
- *Larix decidua*

Broadleaved Deciduous Forest

- Broadleaved deciduous species dominate (oak, lime, ash, maple, elm and in southern European Russia also beech, chestnut and hornbeam).
- In mountainous areas (Caucasus, Southern Ural, Sikhote-Alin) with significant presence of conifers such as spruce, fir and Korean pine.

— *Quercus robur* (Pedunculate oak)

— *Sorbus aucuparia* (Russian Mountain Ash)

— *Acer platanoides* (Norway maple)

— *Ulmus laevis* (Russian elm)

— *Castanea dentata* (American chestnut)

Stone Birch Forest

- Stone birch dominates, often with presence of larch trees or patches of trees.
- In Kamchatka this forest has undergrowth of mountain pine, and in the mountains of Primorye and Sakhalin with presence of spruce and fir.
- Dominant Species - Paper Birch (*Betula papyrifera*)

Other species

— *Betula alleghaniensis* — yellow birch

— *Betula cordifolia* — mountain paper birch

— *Betula lenta* - black birch

— *Betula nigra* - river birch

— *Betula occidentalis* - red birch

— *Betula populifolia* - gray birch

— *Betula pumila - swamp* birch

Dwarf Pine Forest

- Dwarf pine dominates in patches or shrubby forest, often with a sparse upper storey of larch or stone birch.

Dwarf Pine

— *Pinus strobus* - Minuta Pine

— *Pinus parviflora* - Adcock's Dwarf

— *Pinus strobus* - Nana Pine

Other species

— *Betula populifolia* (Gray birch)

— *Betula papyrifera* (Paper Birch)

— *Betula occidentalis* (Red birch)

— *Betula alleghaniensis* (Yellow birch)

— *Larix sibirica* (Russian larch)

— *Larix occidentalis*

Birch-Aspen and Mixed Forest

- Birch, aspen and gray alder dominate, with presence of coniferous trees or patches of trees.
- In most cases, this forest follows logging, clearing or forest fires.
- Important species
  — *Populus tremuloides* (Aspen)
  — *Larix decidua* (European larch)
  — *Betula alleghaniensis* (Yellow birch)
  — *Alnus incana* (grey alder or speckled alder)

Areas of Potential Forest

- Agricultural and other non-forest ecosystems in which climate and soils are suitable for forest growth.

FOREST RSOURCES OF CHINA

- China is the world's third largest country, after Russia and Canada with an area of 9.6 million square kilometers.
- The forestland is the land predominantly used for cultivation, restoration and development of forest vegetation.

China - Forest

Forest as a % of land area	:	18.21%
Forest land area	:	284.91 m ha 704.03 m acres
Total forest area	:	174.91 m ha 426.97 m acres
Forest per capita	:	0.128 ha
Natural Forest area	:	115.8 m ha
Plantation Forest area	:	53.3 m ha

Forestland Area

- On the basis of tree cover and practical land use, forestland is categorized as forested land, open forest land, shrub land, unestablished forest land, nursery land, non-stocked forestland, land suitable for forest and other forestland. The forestland area totals 310.46 million ha.

Natural forest

- The area of natural forest in China is 121.84 million ha, taking up 64 % of forested land.
- The natural forest is mainly distributed in the provinces of northeast and southwest China, among which the natural forest area of 7 provinces including Heilongjiang, Inner Mongolia, Yunnan, Sichuan, Tibet, Jiangxi and Jilin

Plantation Forestry in China

- The top ten tree species (groups) in plantation are

  — *Cunninghamia lanceolata*
  — *Larix gmelinii*
  — *Pinus massoniana*
  — *Pinus tabuleaformis*
  — *Cupressus funebris*
  — *Pinus elliottii*
  — *Robinia pseudoacacia*
  — *Populus* spp.
  — *Eucalyptus* spp.
  — *Quercus* spp.

FOREST TYPES OF CHINA

1. Cold-Temperate and Temperate Mountain Coniferous Forest
2. Temperate Coniferous Forest
3. Subtropical Coniferous Forest
4. Mixed Coniferous and Broad-Leaved Forest
5. Temperate Deciduous Broad-Leaved Forest
6. Subtropical Evergreen Broad-Leaved Forest
7. Tropical Rain Forest/Monsoon Forest
8. Bamboo Forest

Cold-Temperate and Temperate Mountain Coniferous Forest

- Juniper (*Sabina przewalskii*)
- Douglas fir (*Pseudotsuga menziesii*)
- Sitka spruce (*Picea sitchensis*)
- *Picea crassifolia*
- *Cunninghamia* spp.

Temperate Coniferous Forest

- Coast Redwood (*Sequoia sempervirens*)
- Giant Sequoia (*Sequoiadendron giganteum*)

- Douglas fir (*Pseudotsuga menziesii*)
- Sitka spruce (*Picea sitchensis*)
- Alerce (*Fitzroya cupressoides*)
- Kauri (*Agathis australis*)

Subtropical Coniferous Forest

- *Pinus armandii* - Chinese White Pine
- *Pinus bhutanica* - Bhutan White Pine
- *Pinus densata* - Sikang Pine
- *Pinus kesiya* - Khasi Pine
- *Pinus roxburghii* - Chir Pine
- *Pinus tabuliformis* - Chinese Red Pine
- *Pinus wallichiana* - Blue Pine or Bhutan Pine
- *Pinus banksiana* - Jack Pine
- *Pinus clausa* - Sand Pine
- *Pinus echinata* - Shortleaf Pine
- *Pinus glabra* - Spruce Pine
- *Pinus strobus* - Eastern White Pine
- *Pinus taeda* - Loblolly Pine

Other important species

- Chinese Cryptocarya (*Cryptocarya chinensis*)
- *Castanopsis hystrix*
- Japanese blue oak (*Quercus glauca*)
- Formosan alder (*Alnus formosana*)
- Maple (*Acer* spp.)
- Taiwan hemlock (*Tsuga chinensis*)
- Spruces (*Picea* spp.)
- Firs (*Abies* spp.)

Mixed Coniferous and Broad-Leaved Forest

Dominant species

- *Pinus armandii - Chinese White Pine*
- *Nothofagus*
- *Eucalyptus*

Other species

- Oaks (*Quercus* spp.)
- Beeches (*Fagus* spp.)
- Maples (*Acer* spp.)
- Birches (*Betula* spp)
- Firs (*Abies* spp.)
- Spruces (*Picea* spp.)

Temperate Deciduous Broad-Leaved Forest

Pines

- Ponderosa Pine (*Pinus ponderosa*)
- Sugar Pine (*Pinus lambertiana*)
- Incense Cedar (*Calocedrus decurrens*)
- White Fir (*Abies concolor*)
- Douglas Fir (*Pseudotsuga menziesii*)
- Giant Sequoias (*Sequoiadendron giganteum*)

Broadleaved trees

- Yellow Larch (*Larix occidentalis*)
- Black Oak (*Quercus kelloggii*)
- Canyon Live Oak (*Quercus chrysolepis*)
- Dogwood (*Cornus* spp.)
- Mountain Misery (*Chamaebatia foliolosa*)
- Manzanitas (*Arctostaphylos* spp.)
- *Alnus nepalensis*
- *Acer oblongum*
- *Bombax malabricum* (Semal)
- *Terminalia termentosa* (Asna)
- *Olea cuspidata*

Subtropical Evergreen Broad-Leaved Forest

- *Alnus nepalensis*
- *Acer oblongum*
- *Bombax malabricum* (Semal)
- *Terminalia termentosa* (Asna)
- *Olea cuspidata*

Tropical Rain Forest/Monsoon Forest

Monsoon forest

- A tropical forest with a marked seasonal rhythm induced by alternating wet and dry seasons typical of the true monsoon lands.

Important species

- Blue pine (*Pinus wallichiana*)
- Spruce (*Picea smithiana*)
- Deodar (*Cedrus deodara*)
- *Cupressus torulosa*
- Silver fir (*Abies pindrow*)
- Birch (*Betula utilis*)
- Boxwood (*Buxus wallichiana*)

Bamboo Forest

- China's famous Wine Capital is Sichuan province, Shunan Bamboo Forest (Shunan Zhuhai) is the oldest and biggest bamboo forest in China, by covering an area of 120 square kilometers, and home to around 400 different types of bamboo. It is considered as a natural oxygen bar.

Dominant Bamboos

— *Bambusa sinospinosa*
— *Dendrocalamus stocksii*
— *Phyllostachys nigra*
— *Melocanna bambusoides*
— *Bambusa sinospinosa*
— *Bambusa ventricosa*
— *Dendrocalamus latiflorus*
— *Chimonobambusa quadrangularis*
— *Chimonocalamus pallens*
— *Phyllostachys edulis*
— *Phyllostachys bambusoides*
— *Phyllostachys sulphurea*
— *Phyllostachys aurea*
— *Phyllostachys nigra*
— *Cephalostachyum pergracile*
— *Oxytenanthera braunii*
— *Fargesia robusta*
— *Fargesia denudata*

FOREST RESOURCES OF JAPAN

- Japan is stratovolcanic archipelago consisting of about 6,852 islands. The four largest islands are Honshu, Hokkaido, Kyushu and Shikoku make up about ninety-seven percent of Japan's land area and often are referred to as home islands.

Salient features of Japanese Forest

- Japan is one of the world's most heavily forested countries.
- Before humans began altering the landscape, almost all the Japanese islands were covered by forest.
- Today there are 62.3 million acres of forest in Japan, covering two-thirds of the country's mountainous land area.
- Japan has a mild climate and plentiful rain which are ideal for forest growth.
- With the exception of some extreme habitats such as alpine meadows and areas of river gravel Japan's natural cover is almost always forest.

Forest Types of Japan

Tropical broadleaf forest

- Japanese maple (*Acer palmatum*)
- Japanese zelkova (*Zelkova serrata*)
- Japanese black pine (*Pinus thunbergii*)
- Japanese white pine (*Pinus parviflora*)
- Japanese tree lilac (*Syringa reticulata*)
- Cherry trees (*Prunus serrulata*)
- Japanese yew (*Taxus cuspidata*)
- Japanese cypress (*Chamaecyparis obtusa*)
- Japanese persimmon (*Diospyros kaki*)
- Japanese Chinquapin (*Castanopsis cuspidata*)

Temperate broadleaf forest

- Stone pine (*Pinus pumila*)
- Birch (*Betula ermanii*)
- Alder (*Alnus aximowiczii*)
- Mountain ash (*Sorbus sambucifolia*)

Tropical coniferous forest

- Japanese cedar (*Cryptomeria japonica*)
- Japanese cypress (*Chamaecyparis obtusa*)
- Mongolian oak (*Quercus mongolica*)

- Lobed elm (*Ulmus laciniata*)
- Japanese white birch (*Betula platyphylla*)
- Glehn's spruce (*Picea glehnii*)
- Sakhalin fir (*Abies sachalinensis*)
- Japanese black pine (*Pinus thunbergii*)
- Red pine (*Pinus densiflora*)

FAO's Classification of Japanese Forest

There are three main kinds of natural forests in Japan as discussed below

Coniferous forest of spruce and Fir

- Fir (*Abies sachalinensis*)
- Japanese fir (*Abies mariesii*)
- Spruce (*Picea glehnii*)
- Maximowicz spruce (*Picea maximowiczii*)
- Japanese cypress (*Chamaecyparis obtusa*)
- Japanese black pine (*Pinus thunbergii*)

Cool-temperature deciduous forests

- Blue Japanese oak (*Quercus glauca*)
- Japanese Oak (*Quercus mongolica*)
- Japanese beech (*Fagus crenata*)
- Chinese beech (*Fagus engleriana*)
- Japanese maple (*Acer palmatum*)
- Japanese White Birch (*Betula platyphylla*)
- American sweetgum (*Liquidambar styraciflua*)
- Magnolia (*Magnolia liliiflora*)

Broadleaf evergreen forests

- Madrone (*Arbutus menziesii*)
- Magnolia (*Magnolia grandiflora*)
- Golden Chinquapin (*Chrysolepis chrysophylla*)
- Dogwood (*Cornus capitata*)
- Loquat (*Eriobotrya japonica*)

- Longstalk holly (*Ilex pedunculosa*)
- Bay tree (*Laurus nobilis*)
- Glossy privet (*Ligustrum lucidum*)
- Tanoak (*Lithocarpus densiflorus*)

More localized and less widespread forests

1. Subtropical rain forests with large trees, large ferns and strangler vines in the Amami-Oshima Islands and Okinawa Prefecture.
2. Coastal forests, with trees that thrive under windy conditions.
3. Mangrove forests found in some places in the islands south of Kyushu in Okinawa prefecture.
4. Riparian forests, found along mountains streams.
5. Sub-alpine forests with ground hugging trees like dwarf pines and stunted birch that do well in cold, windswept places.

■■■■

CHAPTER - 5

THE STATE OF INDIA'S FOREST

The state of forest report in India has been prepared once in every 2 years by Forest Survey of India, Dehradun. Under the Ministry of Environment, Forest and Climate change the first state of forest report was published by FSI in 1987 and the recent state of forest report has been published in the year 2017. This forest cover assessment provides information on the distribution of forest and tree cover in each and every state and also provides information on the growing stock, carbon stock, and the distribution of trees, bamboos and mangroves. This forest cover assessment also incorporated trees outside forest (TOF). This chapter presents the state of Indian forests under the following sub headings.

- Forest cover measurement
- State wise forest distribution
- Growing stock of Indian forest

FOREST COVER MEASUREMENT

Forest cover: Forest cover is defined as all lands, more than one hectare in area, with a tree canopy density of more than 10%. (Such lands may or may not be statutorily notified as forest area).

The first satellite recorded forest coverage data for India became available in 1987. India and the United States cooperated in 2013, using Landsat MSS with spatial resolution of 80 metres, to get accurate forest distribution data. India thereafter switched to digital image and advanced satellites with 23 metres resolution and software processing of images to get more refined data on forest quantity and forest quality. India now assesses its forest distribution data biannually.

The **Forest Survey of India** classifies forest cover into 4 classes:

S.No	Forest Cover Types	Basic Criteria
1.	**Very dense forest:**	All lands with tree cover (including mangrove cover) of canopy density of 70% and above
2.	**Moderately dense forest**	All lands with tree cover (including mangrove cover) of canopy density between 40% and 70%

S.No	Forest Cover Types	Basic Criteria
3.	**Open forests**	All lands with tree cover (including mangrove cover) of canopy density between 10% and 40%.
4.	**Scrubs**	All forest lands with poor tree growth mainly of small or stunted trees having canopy density less than 10%.
5.	**Non forest**	Lands not included in any of the above classes.

The moderately dense and very dense forests are called Prime forests. Non-Forests are the lands not included in any of the above classes.

Forest Cover in India

According to the FSI (2017), the forest cover in India is estimated to 21.39% of the total geographical area of the country. Out of the total forest cover, the maximum share is of moderate dense forests, followed by Open Forests. Madhya Pradesh has the largest total forest cover, followed by Arunachal Pradesh and Chhattisgarh. Mizoram, has the highest forest cover in percentage terms, followed by Lakshadweep.

Forest cover type	Area (%)
Non-Forest	77.40
Scrub	1.26
Open Forest	9.14
Moderate Dense Forest	9.59
Very Dense forest	2.61

(FSI, 2017)

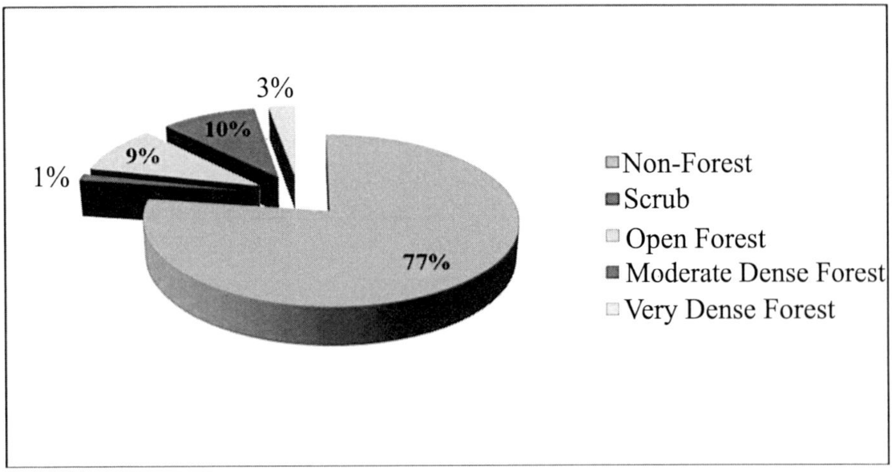

(FSI, 2017)

STATE WISE FOREST DISTRIBUTION

The FSI has reported 70803 sq.km as the total forest cover as per its 2017 assessment and indicated that this forest cover contributes 21.54% of total geographical area of country. This forest cover has reported that forest area of all 25 states and 7 union territories.

The assessment indicated that Madhya Pradesh recorded the largest forest cover 74,414 sq.km followed by Arunachal Pradesh (66,946 sq.km), Chhattisgarh (55,947 sq.km), Orissa (54,345 sq.km) and Maharashtra (50,682 sq.km). The assessment also indicated that seven states viz., Lakshadweep (90.33%), Mizoram (86.27%), Andaman and Nicobar Islands (81.73%), Arunachal Pradesh (79.96%), Manipur (77.69%), Meghalaya (76.45%) and Nagaland (75.33%) recorded more than 75% forest cover. In the remaining 25 states and union territories eight states *viz.,* Assam, Chhattisgarh, Goa, Kerala, Sikkim, Tripura, Uttarakhand, Dadra and Nagar haveli recorded more than 33% forest covers. Other states and union territories have registered low forest cover than the mandated forest cover of 33%.

Compared to 2015 Forest cover assessment the current assessment has recorded an increased forest cover of 6678 sq.km which recorded 0.21% of the total geographic area. The increase in forest cover is maximum in the states of Andhra Pradesh (2141 sq.km), Karnataka (1101 sq.km) and Kerala (1043 sq.km). States like Mizoram, Nagaland, Arunachal Pradesh, Meghalaya and Tripura recorded decline in forest cover.

Table 1. Comparison of State wise forest Cover assessment 2015 and 2017

States/UTs	Geographic Area (sq km)	Geographic Area (sq km)	2015 Assessment		2017 Assessment		Changes	
			Forest Area (sq km)	Forest Cover (%)	Forest Area (sq km)	Forest Cover (%)	Change Area (sq km)	Change Percent (%)
Andhra Pradesh	160204	162968	24424	15.25	28147	17.27	2141	1.31
Arunachal Pradesh	83743		67248	80.30	66964	79.96	-190	-0.23
Assam	78438		27623	35.22	28105	35.83	567	0.72
Bihar	94163		7288	7.74	7299	7.75	45	0.05
Chhattisgarh	135191		55586	41.12	55547	41.09	-12	-0.01
Delhi	1483		188.77	12.73	192.41	12.97	3.64	0.25
Goa	3702		2224	60.08	2229	60.21	19	0.51
Gujarat	196022	196244	14660	7.48	14757	7.52	47	0.02
Haryana	44212		1584	3.58	1588	3.59	8	0.02

States/UTs	Geographic Area (sq km)	Geographic Area (sq km)	2015 Assessment		2017 Assessment		Changes	
			Forest Area (sq km)	Forest Cover (%)	Forest Area (sq km)	Forest Cover (%)	Change Area (sq km)	Change Percent (%)
Himachal Pradesh	55673		14696	26.40	15100	27.12	393	0.71
Jammu & Kashmir	222236		22988	10.34	23241	10.46	253	0.71
Jharkhand	79714	79716	23478	29.45	23553	29.55	29	0.04
Karnataka	191791		36421	18.99	37550	19.58	1101	0.57
Kerala	38863	38852	19239	49.50	20321	52.30	1043	2.68
Madhya Pradesh	308245	308252	77462	25.13	77414	25.11	-12	0.00
Maharashtra	307713		50628	16.45	50682	16.47	-17	-0.01
Manipur	22327		16994	76.11	17346	77.69	263	1.18
Meghalaya	22429		17217	76.76	17146	76.45	-116	-0.52
Mizoram	21081		18748	88.93	18186	86.27	-531	-2.52
Nagaland	16579		12966	78.21	12489	75.33	-450	-2.71
Odisha	155707		50354	32.34	51345	32.98	885	0.57
Punjab	50362		1771	3.52	1837	3.65	66	0.13
Rajasthan	342239		16171	4.73	16572	4.84	466	0.14
Sikkim	7096		3357	47.31	3344	47.13	-9	-0.13
Tamil Nadu	130058	130060	26345	20.26	26281	20.21	73	0.06
Telangana	114865	112077	21591	18.80	20419	18.22	565	0.50
Tripura	10486		7811	74.49	7726	73.68	-164	-1.56
Uttar Pradesh	240928		14461	6.00	14679	6.09	278	0.12
Uttarakhand	53483		24240	45.32	24295	45.43	23	0.04
West Bengal	88752		16828	18.96	16847	18.98	21	0.02
A & N Islands	8249		6751	81.84	6742	81.73	-9	-0.11
Chandigarh	114		22.03	19.32	21.56	18.91	-0.10	-0.09
Dadra & Nagar Haveli	491		206	41.96	207	42.16	1	0.20
Daman & Diu	112	111	19.61	17.51	20.49	18.46	0.88	0.79
Lakshadweep	32	30	27.06	84.56	27.10	90.33	0.04	0.13
Puducherry	480	490	55.38	11.54	53.67	10.95	-3.28	-0.67
Grand Total	3287263	3287469	701673	21.34	708273	21.54	6778	0.21

(FSI, 2015, 2017)

State Wise Tree Cover Distribution

Similar to forest cover assessment, the tree cover assessment of the states and union territories have been furnished in the FSI state of forest report 2017 and is furnished in Table 2.

The assessment indicated that Maharashtra has recorded maximum tree cover of 9831 sq.km followed by Rajasthan (8266 sq.km), Madhya Pradesh (8073 sq km) and Gujarat (8024 sq.km). In case of percent tree cover assessment Daman and Diu recorded the highest tree cover percent of 9.01% followed by Chandigarh (8.77%), Goa (8.73%), Andaman Nicobar islands recorded lowest tree cover assessment of 0.42%

Table 2. Comparison of State wise tree Cover assessment 2015 and 2017

States/UTs	Geographic Area (sq km)	Geographic Area (sq km)	2015 Assessment		2017 Assessment	
			Area (sq km)	Tree Cover (%)	Area (sq km)	Tree Cover (%)
Andhra Pradesh	160204	162968	3965	2.48	3753	2.30
Arunachal Pradesh	83743		761	0.91	807	0.96
Assam	78438		1613	2.06	1496	1.91
Bihar	94163		2182	2.32	2263	2.40
Chhattisgarh	135191		3629	2.68	3833	2.84
Delhi	1483		111	7.49	113	7.62
Goa	3702		325	8.78	323	8.73
Gujarat	196022	196244	7914	4.04	8024	4.09
Haryana	44212		1355	3.07	1415	3.20
Himachal Pradesh	55673		757	1.36	822	1.48
Jammu & Kashmir	222236		8354	3.76	7815	3.52
Jharkhand	79714	79716	2783	3.49	2922	3.67
Karnataka	191791		5552	2.89	5713	2.98
Kerala	38863	38852	2951	7.59	2959	7.62
Madhya Pradesh	308245	308252	7773	2.52	8073	2.62
Maharashtra	307713		9558	3.11	9831	3.19
Manipur	22327		243	1.09	220	0.99

States/UTs	Geographic Area (sq km)	Geographic Area (sq km)	2015 Assessment		2017 Assessment	
			Area (sq km)	Tree Cover (%)	Area (sq km)	Tree Cover (%)
Meghalaya	22429		710	3.17	657	2.93
Mizoram	21081		535	2.54	467	2.22
Nagaland	16579		381	2.30	379	2.29
Odisha	155707		3986	2.56	3993	2.56
Punjab	50362		1544	3.07	1622	3.22
Rajasthan	342239		8269	2.42	8266	2.42
Sikkim	7096		35	0.50	35	0.49
Tamil Nadu	130058	130060	4505	3.46	4671	3.59
Telangana	114865	112077	2549	2.22	2669	2.38
Tripura	10486		233	2.22	215	2.05
Uttar Pradesh	240928		7044	2.92	7442	3.09
Uttarakhand	53483		752	1.41	767	1.43
West Bengal	88752		2088	2.35	2136	2.41
A&N Islands	8249		37	0.44	35	0.42
Chandigarh	114		9	7.80	10	8.77
Dadra & Nagar Haveli	491		28	5.72	30	6.11
Daman & Diu	112	111	10	9.18	10	9.01
Lakshadweep	32	30	4	12.50	2	6.67
Puducherry	480	490	27	5.63	27	5.51
Grand Total	**3287263**	**3287469**	**92572**	**2.82**	**93815**	**2.85**

(FSI, 2015, 2017)

PRODUCTIVITY POTENTIAL (GROWING STOCK) OF INDIAN FORESTS

The estimates of growing stock present in the forest and trees outside forest were recorded in physiographic regions and state wise.

Physiographic Zone Wise

India has been classified into 14 physiographic zones from western Himalayas into eastern Himalayas, plains of India, Deccan plateau and high lands, Ghats and coastal regions. The total growing stock estimated for the entire physiographic zones is 5822.37 million m^3 which include 4218 million m^3 from the forestry sector and 1603 million m^3 from trees outside forest.

The western Himalayas recorded maximum growing stock area of 959.98 million m^3 and the western plains recorded the lowest growing stock of 11.26 million m^3. In case of Trees outside forest, east deccan recorded higher growing stock of 273.88 m^3 and western plains recorded the lowest growing stock of 60.484 million m^3 in trees outside forest. Considering the growing stock of both forest and trees outside forest, western Himalayas recorded increased growing stock of 1149 million m^3. Compared to the growing stock of 2015 estimation, there is an increment of 53.99 million m^3 of growing stock (Table 3).

State Wise Growing Stock Assessment

The growing stock estimations of 25 states and seven union territories are furnished in table 4. Among states and union territories, the maximum growing stock of 420.79 million m^3 is recorded in the state of Arunachal Pradesh followed by Uttarakhand and Karnataka which recorded the growing stock at 418.33 million m^3 and 327.660 million m^3 respectively. Among the states and union territories, Lakshadweep, Daman and Diu, Pudhucherry, Chandigarh and Delhi records less than 1 million m^3 of growing stock.

In case of trees outside forest, Maharashtra recorded increased growing stock of 164.546 million m^3 followed by Jammu and Kashmir (146.082 million m^3). Considering the growing stock available both in forest and trees outside forest, Arunachal Pradesh recorded higher growing stock of 511.488 million m^3. Chandigarh, Daman and Diu, Lakshadweep and Pudhucherry recorded lower growing stock of less than 1 mill. m^3

Productivity Potential (Growing Stock) of Indian Forest

Table 3. Physiographic Zone wise Growing Stock Estimates

Physiographic Zone	Areas of Phy. Zone (sq km)	Growing Stock (million cu m) (2017)		
		Forest	TOF	Total
Western Himalayas	329255	959.980	189.504	1149.484
Eastern Himalayas	74618	391.600	82.256	473.856
North East Ranges	133990	287.240	114.105	401.345
Northern Plains	295780	141.070	108.910	249.980
Eastern Plains	223339	178.310	76.847	255.157
Western Plains	319098	11.260	60.484	71.744
Central Highlands	373675	135.880	118.273	254.153

Physiographic Zone	Areas of Phy. Zone (sq km)	Growing Stock (million cu m) (2017)		
		Forest	TOF	Total
North Deccan	355988	293.740	106.076	399.816
East Deccan	336289	671.540	237.884	909.424
South Deccan	292416	219.270	112.733	332.003
Western Ghats	72381	423.120	117.656	540.776
Eastern Ghats	191698	272.220	66.041	338.261
West Coast	121242	150.200	136.752	286.952
East Coast	167494	82.950	76.476	159.426
TOTAL	3287263	4218.38	1603.99	5822.37
Growing stock (ISFR 2013, 2015)	3287263	4195.047	1573.340	5768.387
Change		23.333	30.657	53.990

(FSI, 2015, 2017)

Table 4. State wise growing stock estimates

States/UTs	Geographical Area (sq km)	RFA (sq km)	Volume of Growing Stock (million cum) (2017)		
			Forest	TOF	Total
Andhra Pradesh	160,204	37,258	156.038	62.348	218.386
Arunachal Pradesh	83,743	51,407	420.793	90.695	511.488
Assam	78,438	26,832	133.128	30.195	163.323
Bihar	94,163	6,493	28.500	37.471	65.991
Chhattisgarh	135,191	59,772	323.660	86.348	410.008
Delhi	1,483	102	0.510	1.274	1.784
Goa	3,702	1,225	9.516	3.832	13.348
Gujarat	196,022	21,647	52.030	113.992	166.022
Haryana	44,212	1,559	5.405	15.490	20.895
Himachal Pradesh	55,673	37033	315.580	23.199	338.779
Jammu & Kashmir	222,236	20,230	232.837	146.082	378.919
Jharkhand	79,714	23,605	117.085	64.391	181.476

States/UTs	Geographical Area (sq km)	RFA (sq km)	Volume of Growing Stock (million cum) (2017)		
			Forest	TOF	Total
Karnataka	191,791	38,284	327.660	89.585	417.245
Kerala	38,863	11,309	167.085	51.898	218.983
Madhya Pradesh	308,245	94,689	285.611	97.474	383.085
Maharashtra	307,713	61,579	252.572	164.546	417.118
Manipur	22,327	17,418	53.200	7.750	60.950
Meghalaya	22,429	9,496	38.720	17.254	55.974
Mizoram	21,081	5,641	19.052	42.762	61.814
Nagaland	16,579	9,222	36.899	11.425	48.324
Odisha	155,707	58,136	258.013	81.503	339.516
Punjab	50,362	3,084	12.941	20.012	32.953
Rajasthan	342,239	3,2737	46.274	81.865	128.139
Sikkim	7,096	5,841	26.280	1.951	28.231
Tamil Nadu	130,058	22,877	127.810	66.257	194.067
Telangana	114,865	26,904	60.462	39.660	100.122
Tripura	10,486	6,294	21.740	6.498	28.238
Uttar Pradesh	240,928	16,582	135.848	88.340	224.188
Uttarakhand	53,483	38,000	418.332	19.855	438.187
West Bengal	88,752	11,879	76.634	38.100	114.734
Andaman & Nicobar Islands	8,249	7,171	55.928	0.584	56.512
Chandigarh	114	35	0.321	0.103	0.424
Dadra & Nagar Haveli	491	204	1.850	0.785	2.635
Daman & Diu	112	8	0.001	0.115	0.116
Lakshadweep	32	0.00	0.000	0.060	0.060
Puducherry	480	13	0.065	0.298	0.363
Total	**3287263**	**764566**	**4218.380**	**1603.997**	**5822.377**

(FSI, 2015, 2017)

CHAPTER - 6

SILVICULTURE – PRINCIPLES AND PRACTICES

SILVICS AND SILVICULTURE

Silviculture has been defined variously by different authors.

- 'Silviculture is that branch of forestry which deals with the establishment, development, care and reproduction of stands of timber'.
- 'The art and science of cultivating forest crops'.
- 'The term silviculture in English commonly refers only to certain aspects of theory and practice of raising forest crops'.

In a broad sense, silviculture may be taken to include both silvics and its practical application.

According to Indian Forest and Forest Products Terminology, silvics is 'the study of life history and general characteristics of forest trees and crops with particular reference to environmental factors, as the basis for the practice of silviculture'.

Objectives of Silviculture

(i) Production of species of economic and commercial value

(ii) Production of higher volume per unit area

(iii) Production of quality timber from desired species

(iv) Reduction of rotation

(v) Raising forests in blank areas, wastelands and other unutilised lands

(vi) Creation of manmade forests in place of natural forests

(vii) Introduction of exotics to meet the ward requirement.

Difference between Silvics and Silviculture

Silviculture	Silvics
Silviculture is the branch of forestry which deals with the establishment, development, care and reproduction of stands of timber	silvics is the study of life history and general characteristics of forest trees and crops with particular reference to environmental factors
It is the science dealing with both theory and practices	It is the basic theory upon which silviculture stands
It is a broder term to cover all aspects of forestry	It is a part of silviculture
It deal with the procedure of obtaining natural regeneration under various systems and methods of tending of young crops	It deals with the study of tree and forests as biological units, laws governing the growth and effect of environment on vegetation
It is application of silvics	It is the principle for siliculture

Regeneration – Natural and Artificial

Regeneration is a synonym for reproduction in which the tree species reproduce new individuals of their own species either naturally or artificially. The term regeneration is originally used for natural process but later on the term regeneration extended to other methods of propagation too. Regeneration is predominantly by the following methods.

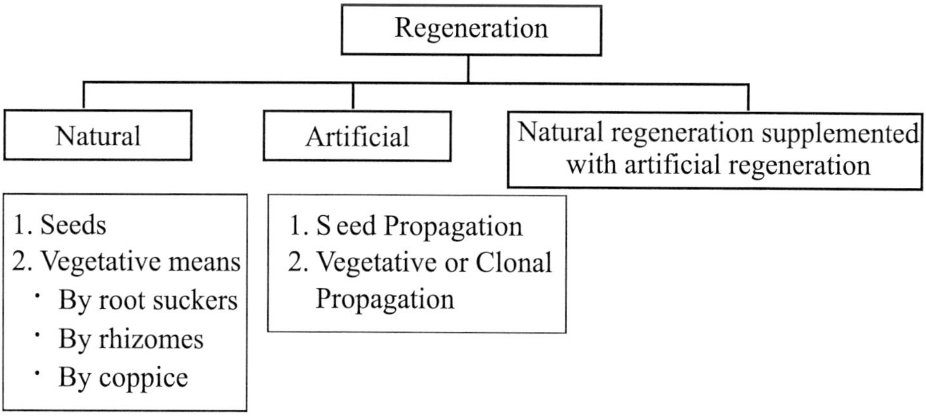

Natural Regeneration

It is a regeneration process in which the renewal of a tree crop is by natural means of seed sowing, germination and establishment or by means of coppice shoots or through root suckers.

Example: The regeneration process that occurs predominantly in dry and moist deciduous forests.

Natural Regeneration Methods

Seed based Natural Regeneration

Natural regeneration by seed is the most robust way of regeneration in many forest areas. It is usually through seeds and depends on the following factors.

(i) Seed production and dispersal

Seed production in a tree species should be adequate to ensure successful natural regeneration. The quantity and quality of seed depends upon the species, age of the trees, crown size, apical dominance and the associated biotic and abiotic factors. All the tree species do not produce seeds annually.

Based on seed production either annually or periodically, years are classified into good seed year, moderate seed year or bad seed year. Natural regeneration is very good in good seed year and very poor in bad seed year. Hence, according to species and seed production capacity, natural regeneration varies from site to site and species to species.

Not only seed production is essential but also its dispersal to ideal locations is significantly important for the success of natural regeneration. Good seed production in wrong locality and bad seed production in good locality may lead to poor regeneration. The seeds produced have to be dispersed safely by all means to ensure and complete the natural regeneration process. Examples of seed dispersal

(i) Wind (*Casuarina, Terminalia, Poplar, Pterocarpus, Acacia, Dipterocarpus* etc.)

(ii) Water (most mangroves and riverine species)

(iii) Gravity (Dispersal by weight - *Quercus, Juglans, Aesculus* etc.)

(iv) Birds (*Prunus, Mulberry, Diospyros* etc.)

(v) Animals (*Prosopis, Acacia* etc.).

(ii) Seed germination

Successful natural regeneration is possible only with profuse seed germination. However, germination depends on internal and external factors. The internal factors include permeability of seeds to water and air, embryo development, embryo maturity, viability and germinability of the seeds. Tree species differ significantly for all the above factors. The external factors include all biotic and abiotic factors of the locality which include light, temperature, humidity and site quality.

(iii) Seedling establishment

Germinated seeds should develop into a seedling for successful completion of the regeneration process. Seedling establishment is very essential and it depends on the favourable growing conditions in a locality. The growing seedlings have to be protected from the biotic and abiotic components of the locality.

(iv) Genetic factors

Good natural regeneration could be achieved only when the seeds are produced from genetically superior trees. Trees which are genetically superior usually produce quality seeds and thus ensure successful natural regeneration resulting in good natural forests. Seed production in original geographical locations/land races/centers of origin has a significant impact on the natural regeneration.

(v) Soil factor

Natural regeneration is very successful in good quality site, free from weeds, leaf litter, excessively moist or dryness and other inhibiting factors. Natural regeneration is more in clear felled forest areas compared to the abandoned sites due to richness in soil organic matter and associated humus accumulation, high porosity, water holding capacity and adequate soil microflora and microfauna. Problem soils like acid, saline and alkaline are not suitable for many tree species. In such soils, soil working processes improve seed germination. A soil with good drainage usually contributes more towards seed germination compared to clay or waterlogged soils.

(vi) Climatic factors

Light, temperature and humidity of a locality play a significant role in natural regeneration. Availability of adequate light intensity ensures good germination. Many a time, light reaching the forest floor is influenced by the canopy growth and density which however, can be manipulated to allow the much needed light for natural regeneration. Light also controls the temperature and the soil moisture content of the locality. Too much of light leads to high temperature and causes rapid drying of the soil moisture and results in poor germination and regeneration. Light demanding species usually germinate very easily in such conditions compared to shade bearers and shade demanders. Frost and cold injury are also serious inhibiting factors in certain forest types. Frost and cold sensitive species may not have natural regeneration under abnormal climatic conditions. The severity and longevity of frost and snow also damage the regeneration and affect the whole regeneration process.

(vii) Biotic factors

Natural regeneration is effective when the regeneration areas are protected from all biotic factors. Wild animals particularly deer, porcupine, bisons, wild pigs etc. damage the young regeneration and affect the whole natural regeneration process. Similarly, damage by pests and diseases also lead to poor natural regeneration.

Natural Regeneration by Vegetative Means

Natural regeneration also occurs through vegetative plant parts in some species as indicated below.

(i) Natural regeneration by coppice

Certain species express multiplication through coppice. It is a method of vegetative reproduction, in which the cut end of the stems of saplings, poles and trees produces shoot buds which ultimately lead to the development of a matured tree.

E.g. Eucalyptus, Casuarina, Melia, Teak, Ailanthus etc.

Coppice multiplication depends on the season, size of the stool, age of the stool, silvicultural system, coppicing ability, rotation followed etc.

Based on the nature of the coppice, natural regeneration is classified into two types-

(a) Seedling coppice

This occurs mainly in seedlings/saplings when they are cut at the base which will induce adventitious buds that leads to multiple shoots. This method is common in natural forests and this seedling coppice is currently used in vege-tative multiplication of potential clones by exploiting the apical dominance of coppice shoots.

(b) Stool coppice

A stool is a stump left after felling the tree which is usually around 15 cm height from the ground level. The cut portion of the stool induces adventitious buds which may lead to multiple shoots and they are referred as stool coppice. From this coppice shoots, 2 or 3 leading shoots are allowed to grow into a mature tree. Again, they are cut at the base and the stools are again regenerated into coppice shoots and the whole process is repeated. These coppice shoots are now artificially managed through fertigation to augment the growth of coppice shoots and to reduce the rotation.

Example: Eucalyptus coppiced for three rotations, *Melia dubia* coppiced for two rotations.

Based on the coppicing ability, tree species are classified into

(a) **Strong coppicer** – Eucalyptus, Teak, Shorea, *Dalbergia, Albizzia, Cassia, Prosopis, Melia, Anogeissus*, *Morus alba* etc.

(b) **Fair coppicer** – *Terminalia, Pterocarpus, Quercus, Hardwickia, Chloroxylon*

(c) **Bad coppicer** – *Adina cordifolia, Madhuca indica, Bombax ceiba,* Casuarina.

(d) **Do not coppice** – *Pinus, Picea, Abies, Cedrus*

(ii) Natural regeneration by root suckers

Root suckers are naturally induced from the adventitious buds formed on any part of the root system. Normally, these root suckers are induced due to stress or injury to the roots. These root suckers grow and develop into a full tree. However, the shoots developed are weaker and susceptible to wind fall due to poor growth rate.

Example: Dalbergia sissoo, Dalbergia latifolia, Bombax ceiba, Diospyros melanoxylon, Anogeissus latifolia, Melia dubia etc.

(iii) Natural regeneration by rhizomes

Rhizome is the underground and modified root which will develop into a whole plant.

Example: Bamboos.

Artificial regeneration

The renewal of a forest crop artificially by sowing seeds or by planting seedlings/ cuttings/clones is referred as artificial regeneration. Artificial regeneration is the major method of regeneration in plantations wherein the forest crop is raised / regenerated artificially either by sowing or planting.

Example: Plantations of teak, Sal, Gmelina, Red Sanders etc.

Objectives of artificial regeneration

- To assist or supplement natural regeneration
- To change the composition of the existing tree crop.
- To develop plantations for domestic and industrial utility.
- To rehabilitate barren and wastelands.
- To develop agro and farm forestry plantations.
- To introduce exotics.

Methods of artificial regeneration

Artificial regeneration is carried out by two methods

Seed propagation

In this method, propagation is through seeds collected from selected superior trees or from seed orchards. Seed propagation is one of the easiest methods for all species and it is less expensive, easily manageable and for many species it is the only method of propagation. Seed propagation can be done by direct sowing as in *Acacia nilotica, Adina cordifolia, Dalbergia sissoo, Prosopis juliflora* etc.

For other species, seeds are sown in nursery bed and transplanted into polythene or other containers to develop into seedlings. For some species, seeds can be sown directly in the polythene containers for germination. These seedlings on attaining maturity will be deployed for regeneration to meet the objectives envisaged above.

Example: Teak, Casuarina, Eucalyptus, Neem, Red Sanders, Cassia, Mahua etc.

Advantages of seed propagation

- Easy method of propagation.
- Less expensive and easy management.
- Amenable for all kind of species.
- Successful method for difficult sites.

Disadvantages of seed propagation

- Genetic variability of due to cross pollination and therefore causes variations in plantations resulting in poor productivity and variability in growth.
- Some species express recalcitrant nature in seed germination as in Melia and teak which will result in increased time period to produce seedlings.
- Sterile and parthenocarpy species cannot be multiplied through seeds.

(B) Vegetative propagation

It is otherwise called as asexual propagation in which the propagation is predominantly by vegetative means either through plant parts or through plant tissues.

Advantages

- Vegetative propagation helps to produce genetically identical plants.
- Many tree species do not produce viable seeds and some trees do not produce seeds at all. Such species can be propagated through vegetative means easily.
- Vegetative propagation through grafting helps to produce pest and disease free plants.
- This method eliminates the variability and maintains clonal fidelity.
- It also helps to induce early flowering and reduce the rotation.

- Vegetative propagation helps to accelerate the breeding process and has excellent application in potted or controlled breeding.
- It is more rapid, uniform and large scale multiplication is economically viable compared to seed propagation.
- It helps to preserve the genotype through clonal bank approach.

Disadvantages

- Vegetative propagation is expensive than seed propagation in many species.
- Vegetative propagules lack taproot system and hence prone to damage due to winds and cyclone.
- Chances of epidemic pest and disease occurrence are more compared to seed propagation.

Methods of Propagation

Vegetative propagation is possible by the following two methods.

Macro vegetative propagation

i. Propagation by cuttings

The term cutting refers to any portion removed from the parent plant which is meant for regeneration. The following cuttings are generally employed for vegetative propagation *viz.,* stem cuttings, root cuttings, leaf cuttings, leaf bud cuttings, softwood cuttings, semi hardwood cuttings, hardwood cuttings.

Cuttings are collected from the mother tree or from hedge garden or from mini garden, treated properly to avoid fungal infection and on treatment with growth regulators, the cuttings are planted in containers and kept for rooting under mist chamber conditions.

ii. Propagation by grafting

Grafting is an art of joining plant parts together in such a way that they will readily unite and continue to grow as one plant. A graft has two parts viz., scion and root stock. The scion refers to the upper or top portion of the plant collected from identified mother tree and the roots stock is the lower portion of the graft combination which is raised predominantly through seeds. For successful union of graft, it is very essential that the cambium of scion be placed in close contact with the cambium of the stock.

Grafting Techniques

The following sequential process will facilitate successful grafting.

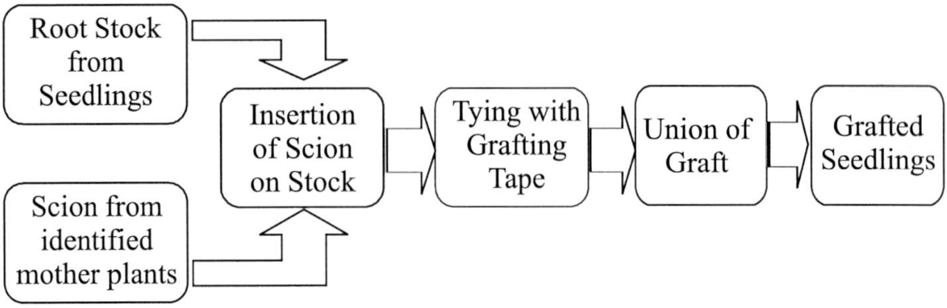

Methods of Grafting

Splice or whip grafting

First the scion and stock plants are selected. Then the stock is cut back to the height where grafting is to be done and a slanting cut of about 3-5 cm long is made and a similar cut of the same length and slope is given to the scion plants which ensures perfect match. These two cut surfaces are placed together and wrapped tightly with polythene grafting tape.

Whip and tongue grafting

It is another grafting method which is applicable when the stock and scion are of similar thickness. In general seedlings, rooted layers or cuttings are whip grafted in late winter or early spring.

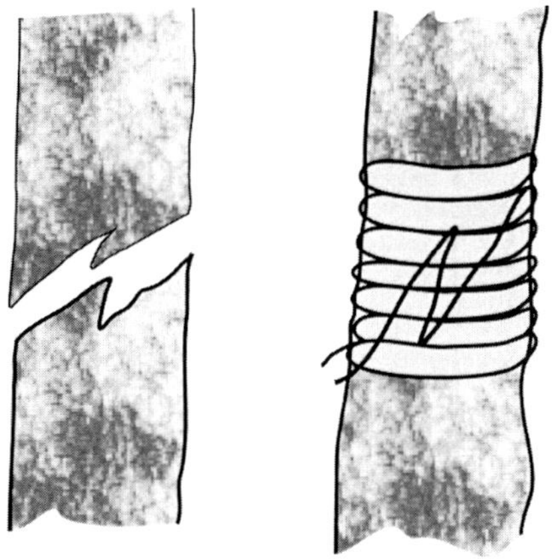

Whip and tongue grafting

The scion and stock are prepared as in the case of whip grafting. On each of these cut surfaces, a reverse cut is made. It is started downward at a point about one third of the distance from the tip, and should be about one-half the length of the first cut. The scion is then slipped into the stock so that they interlock.

- Diagonal cut at stock and scion with tongue
- The graft is tied and wrapped.

Saddle grafting

The stock and the scion should be nearly equal in size for good cambial contact and complete graft. The rootstock is prepared in the form of a wedge. A vertical cut is then given at the base of the scion and is inserted to the rootstock so that they fix well. The components are then tied and coated with grafting wax.

Cleft or Wedge Grafting

It is done in early spring when the tree is nearing dormancy. The shoot tip of the stock plant is removed and a vertical cut is made through the centre of pith. The scion is prepared by making tapering cuts in the form of a wedge. It is then inserted into the split so that the cambial layers of both the scion and the stock are well matched. The union is tied and covered with grafting wax to avoid water entry.

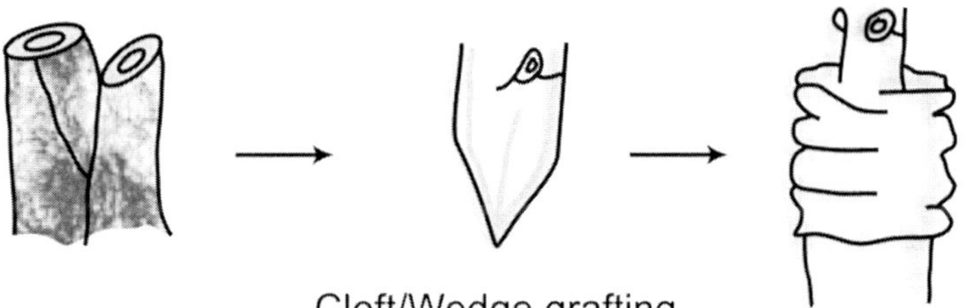

Cleft/Wedge grafting

Side Grafting

An oblique or slanting cut is made in the side of the stock for about 3.5 cm. The scion is prepared with sloping cuts, opposite to each other on the base. Then the rootstock is slightly bent such that it allows sufficient space for scion insertion. The scion should match the cambial layers of the stock. Then the graft is tied with polythene tape to seal the entire cut area. After graft union, the top of the stock is cut and painted with tree paint.

Veneer Grafting

The base of the scion is cut diagonally and a wedge shaped patch of bark of the same size is removed from the side of the stock. The scion and stock are bound together with cambial layers matching. After healing of stock, the upper portion of stock is removed.

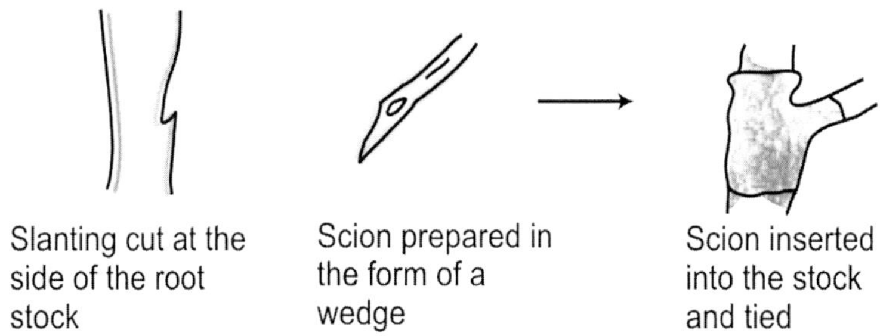

Approach Grafting

Two independent self sustaining plants are grafted together. A cut is made through the cambium and slightly into the wood is given on both stock and the scion. The cut surfaces are pressed firmly and tied with polythene tape. After the union is achieved the scion is severed below the union and the stock above the union.

- Removal of a thin slice of wood and bark from both stock and scion.
- Positioning and tying the stock and scion.
- Waxing the graft union.

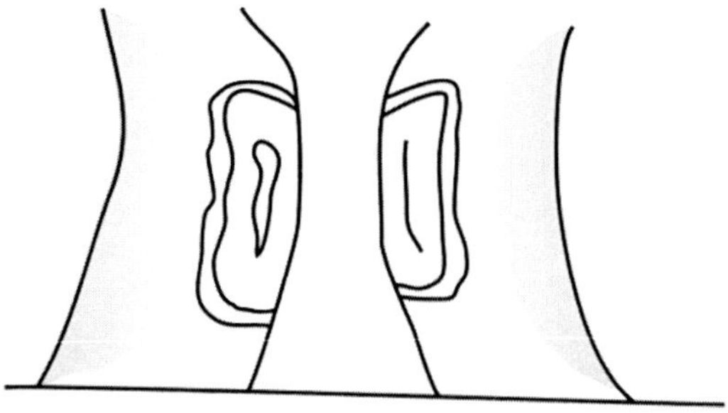

Removal of Bark with a thin slice of wood from the scion and stock

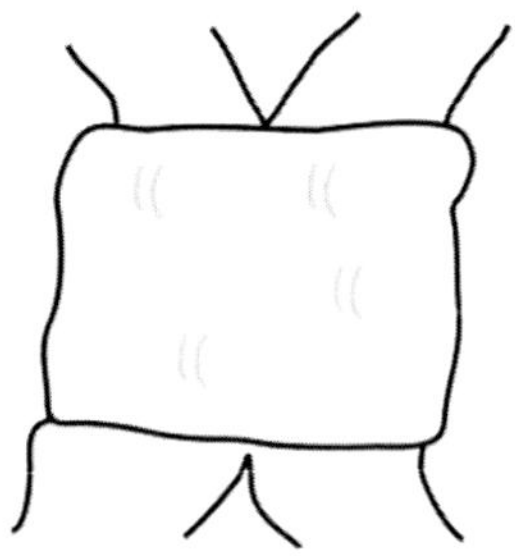

Waxing the graft union

Bark Grafting

It is done in early spring at a time when the bark separates from the wood with ease. In this method, the stock is sawed off and the bark is split downward from the apex about 5 cm long. The scion is inserted between the bark and the wood of the stock by placing longer cut of scion against the cambium. The scion is then tied and covered with grafting wax.

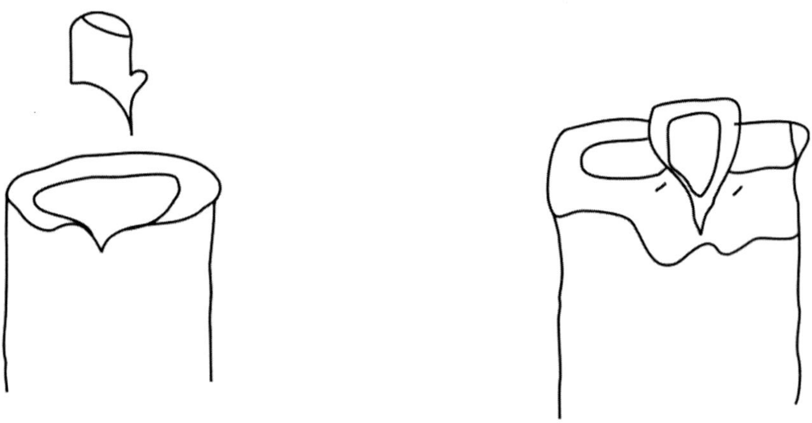

Bark grafting

Root Grafting

The rootstock seedling or rooted cutting or layered plant is dug up, and the roots are used as the stock for the graft. The entire root system may be used, or the roots may be cut into small pieces and each piece is used as a stock. As the roots used are relatively small, whip grafting method is generally followed.

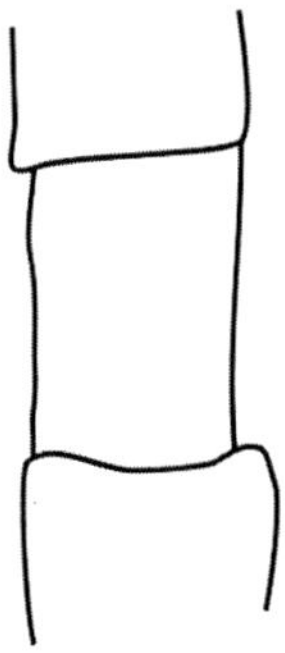

Trimmed Bark in root stocks

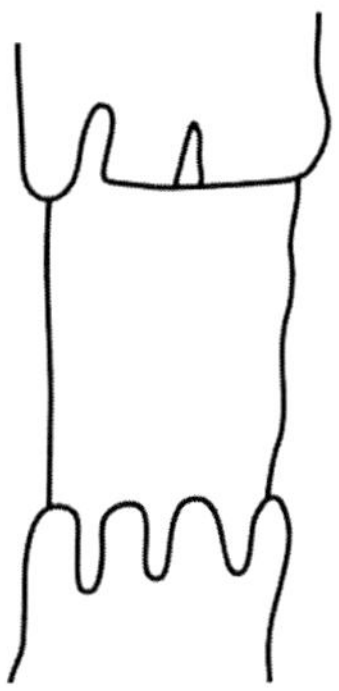

Cuts made in the stock for scion and insertion

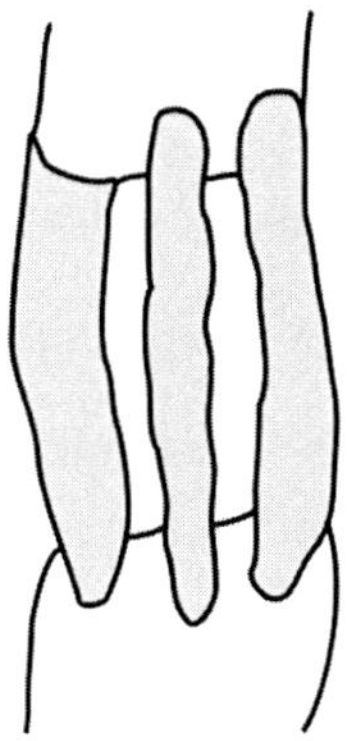

Inserted scion on the root stock Bridge grafting

Bridge Grafting

It is a form of grafting followed when there is injury to the trunk. The wounded area is trimmed. The scion should be taken from dormant one year old shot in the form of strips of uniform size. It is then inserted around the injured section for every 5-7.5 cm and connected to both the upper and lower ends into live undamaged bark.

Among all the grafting methods, as far as forest tree species are concerned, cleft grafting has been found to be more useful.

Propagation by layering

It is an auto vegetative propagation method and is popularly known as Chinese layering or marcottage. This method is practiced in a wide range of tree species due to its simplicity. There are two types of layering viz.

a. **Air layering:** Large shoot of one o two years old are most suitable for making air layers. First the leaves are removed from the selected shoot. Then the stem is girdled by removing a ring of bark about two to three cm wide. The girdled/ cut portion is covered with *sphagnum* moss and wrapped with polythene film tightly to retain the moisture and allows gaseous exchange. This portion will induce roots in four to eight weeks and the rooted portion of the branch is separated out and planted in containers for hardening and further planting in the field.

b. **Ground layering:** In this method, the girdled stem is buried inside the soil. As soon as the rooting occurs, the plant is separated out and planted in containers for hardening and further planting in the field.

Micro vegetative propagation

This is one of the methods of propagation wherein micro shoots viz.; explants (shoot tip, node, internode, meristem, inflorescence, leaf and leaf bits) are *cultured under in-vitro* conditions. Micropropagation leads to mass multiplication from a single explant.

Stages of micropropagation

- Establishment stage
- Shoot multiplication stage
- Root multiplication stage
- Hardening stage

Advantages of micropropagation

- Mass multiplication
- Production of uniform plants
- Production of disease free plants
- *In-vitro* conservation
- Production of secondary metabolites
- Easy exchange of germplasm
- Production of haploids and triploids

Mini Clonal Propagation

A mini clonal technology has been developed for *Casuarina eucalyptus*, *Dalbergia kadaus* and Melia which is one of the pioneering attempts in the country for these industrial wood species. Under this technology, the superior clonal plants were planted in a mini clonal garden and are provided with regular irrigation and fertilization in order to enhance shoot multiplication. In this method, the mother plants are planted at 10 cm × 10 cm spacing and 60 days old plants are ready for collection of cuttings.

Clonal Garden Establishment

The clonal garden can be established at a size of 10 × 1 × 0.6 m or 5 × 1 × 0.6 m or 3 × 1 × 0.6 m using cement trough or GI trough. The bed should be filled with 20 mm stones upto 25 cm and over which finely sewed river sand can be filled. The trough should have facility for drainage. The raised beds can be covered with

100 micron UV stabilized polyethylene film on the top and covered with insect proof mesh to protect the plants and to ensure its freeness from pest and diseases.

Fertigation

The clonal garden should be maintained with irrigation at an interval of every one hour and supplemented with the following nutrients.

- Urea – 300-400 g/m^2
- SSP – 150-175 g/m^2
- KCL – 175-250 g/m^2
- Micro nutrient mixture – 100 g/m^2

The nutrients can be applied twice or thrice depending on the rate of growth of plants.

Clonal Management

The plants are allowed to grow upto 60 days by applying the required nutrient composition. After 60 days the plants are pruned at required size preferably at half of the plant to induce new shoots. With continuous irrigation and nutrient management the cuttings will start producing shoots from 8-10 days onwards and after 15-20 days the cuttings can be collected and treated with 2% carbendazim solution.

Clonal Treatment

The newly induced shoots were separated from the plants and are treated with or without 1500 ppm IBA (liquid formulation) and planted in 90 cc root trainers filled with decomposed coir pith. The rooting started in 15 days and 25 days old rooted plants are ready for hardening.

Green House Conditions

The root trainers are kept under green house conditions with a temperature regime of 32–35°C and a relative humidity of 85-95%. Periodical watering once in every 30 minutes is preferred.

Acclimatization and Hardening

The rooted plants are hardened in a shade house condition with 50% shading for 7-15 days and maintained with adequate irrigations. After hardening chamber, the plants are lifted to open nursery for 30 days. Watering is done 2 times a day and the fertilizer of all 19 (N:P:K) can be applied at the rate of 5 g/plant. During this hardening, application of carbendazim (2 g/l) or triazophos (2ml/lr) is recommended based on the incidence of diseases and pests.

3. NATURAL REGENERATION SUPPLEMENTED WITH ARTIFICIAL REGENERATION

It is a process which is usually practised in forests where natural regeneration is not very successful. Under such conditions, natural regeneration is adequately supported either by sowing or planting through artificial means.

Example: Sal forests in moist deciduous forests of central and North India, teak forests in dry deciduous forests of Madhya Pradesh, Maharashtra and Andhra Pradesh.

NURSERY TECHNOLOGY

Nursery is an area where seedlings are produced or seeds are sown and seedlings are raised for the purpose of planting out for afforestation and reforestation purposes.

Importance of Nursery

- Nursery is very important in the development of plantations both for domestic and commercial utility.
- It helps to produce seedlings of small, medium and tall plants to cater the needs of industrial / social plantations, urban and recreational programmes.
- Nursery helps to produce quality and healthy seedlings
- Nursery grown seedlings ensure higher survival and early establishment.
- Nursery grown seedlings are amenable for both dry and irrigated plantation development programme.
- Nursery grown seedlings are mostly free from pest and diseases and thus help in establishing a hygienic plantation while planting out.

Types of Nursery

Nursery established for tree seedling production plays a vital role in the plantation development programme. Based on the need and the management requirement, tree nurseries are classified into the following types-

(i) Based on longevity/duration of operation

(a) Temporary nurseries

(b) Permanent nurseries

(ii) Based on water availability

(a) Dry nursery

(b) Wet nursery

(iii) Based on the purpose and location

(a) Centralized nursery

(b) Decentralized nursery / Site specific nurseries

Based on Longevity/Duration of Operation

Temporary nurseries

A nursery is said to be a temporary nursery when it is set up for a specific period and to fullfill the seedling requirements of a smaller area.

Temporary nursery is generally shifted from one place to another place depending upon the demand and supply of nursery stock. The seedlings produced in temporary nurseries are mainly used for gap filling or causality replacement and supplementing planting stock. Since the nurseries are put up for only a short period, there are no permanent infrastructure facilitiesand specific staff position. The major advantage of these nurseries is that the cost involved is less.

Since the temporary nurseries are put up at the planting site, seedling transportation is only for a short distance. Temporary nurseries should not at any cost be put up on unfavourable site which lacks water, labour and inspection facilities.

For example: All nurseries available with Forest extension centres are of temporary nature based on the project implemented by the concerned state forest department.

Permanent nurseries

A nursery is said to be a permanent nursery when it supplies seedlings throughout the year/every year to cater the needs of seedling requirement for planting in larger areas and also for a longer period of time. These nurseries are centrally located with permanent infrastructure facilities and staff position in addition to transporting facilities, good water source, labour and equipment facilities. This is generally put up near large plantations with definite felling cycles so as to supply fresh nursery stock during each felling cycle in addition to causality replacement and also during afforestation works when carried out on large areas.

e.g.: Clonal nursery complex, FCRI, Central Plantation nursery, TNPL.

Based on Water Availability

Dry nursery

Dry nurseries are established only for a temporary period without any irrigation. The seeds are processed and sown immediately after the onset of monsoon and seedlings are deployed for direct planting or transplantation purpose. Such nurseries are now becoming rare and obsolete. Present day plantations demand seedlings of good and high quality and dry nurseries hardly have the potential to supply quality seedlings.

Wet nursery

It is a nursery established and maintained with irrigation facility throughout the nursery period. Many of the Government nurseries and most of the private nurseries come under this category.

Based on the Purpose and Location

Centralized nursery

It is a nursery established permanently similar to permanent nursery to provide the large volume of seedling requirement for afforestation, reforestation and agroforestry purposes. As a centralized nursery, all components of infrastructure for both seedling and clonal propagation are established. Most of the wood based industries develop such centralized nurseries to cater the seedling requirement for their own agroforestry and farm forestry plantation development programmes.

e.g.: Centralized nursery established by TNPL and SPB in Tamil Nadu and ITC in Andhra Pradesh.

Decentralized nursery

These nurseries are also permanent in nature but established in specific localities distributed across the state to cater the needs of seedling requirements of such localities. Industries like TNPL and Seshasayee Papers and Board Limited have established decentralized nurseries in association with farmers and other local nursery groups by providing necessary technological support. These nurseries are also provided with quality seeds for seedling production and clonal mother plants for clone production.

e.g.: Decentralized nurseries at Coimbatore, Erode, Karur, Villupuram, Cuddalore etc. established through consortium of industrial agroforestry.

Nursery Location and Physical Requirements

The location of a nursery is very essential to produce quality planting stock coupled with their accessibility to roads and availability of proper communication facilities. The following factors are very essential for locating a nursery.

Location

The area chosen for nursery should be centrally located for economic transport of materials like potting soil, sand, vermiculite, water etc. and easy lifting of nursery stock. The location should be accessible for transportation with good communication facilities including data transfer.

Light

This is an important factor to be considered in selecting an area for raising seedlings. The nursery area should be free from permanent shade. The seedlings should not be raised under tree canopy which cause total shading. Seedlings grown under permanent shade will have lanky growth due to etiolation and therefore, its establishment in the main field would a cause of concern.

Slope

Nursery should be gentle and flat to ensure good drainage facilities for proper growth and establishment of the seedlings. Generally, too sloppy areas can be avoided for nursery establishment.

Size

The size of the nursery should be decided based on the seedling demand, availability of water and other infrastructural facilities. Choosing a large area for a small demand is unmanageable and uneconomical. A minimum nursery site of 1 hectare is preferred but depending on the quantity of seedlings to be produced, the size of the nursery can extend upto 4 hectares (10 acres).

Soil

This is the single and most important factor which decides the quality of the nursery produce. The nursery should be selected on the best available soil preferably on deep sandy loamsoil. In any case, sand is preferable to heavy clay. The soil pH between 6.5 and 7.5 is preferred for production of quality planting materials.

Water

The site chosen for nursery should have a good source of water for watering the seedlings particularly during the driest months. The pH of water should be around 7.0 and the water should be free from high concentrations of carbonates of calcium, magnesium and potassium. Similarly, the water should be free from high concentrations of chlorides of sodium and potassium, sulphate and phosphates of calcium to avoid drying of plants and to ensure production of quality planting material.

Labor

Nursery activities demand huge volume of labour requirement to cater the needs of input transportation, loading and unloading, mother bed formation, sowing, bag filling, transplantation, watering, pest and disease management etc. All these activities demand continuous availability of labour and hence the nursery should be located near to a village or an area where sustainable work force is available for successful establishment and management of the nursery. Nowadays due to mechanization, the labour requirement has been drastically reduced. Nurseries should invest on necessary infrastructure particularly microirrigation facility, mechanized filling operations etc. to manage the labour problem.

Flood

The site should not be located in the areas prone to periodic and recurrent flooding. Sites with heavy clay soil is prone to water logging and flooding and hence this site should be avoided.

Frost

Nursery site should be avoided when the area is subjected to frost and snow.

Wind

Areas subjected to strong winds should also be avoided. Alternatively, such areas when selected for want of other facilities should be protected by establishing one or two rows of windbreaks or a site specific shelterbelt depending on the prevailing wind speed.

Nursery Layout

Once the nursery area is identified and selected, proper planning is essential to establish the nursery. A modern and centralized nursery should have all infrastructure to support seedling production. A typical modern seedling and clonal nursery layout is furnished in Fig. 1 and 2.

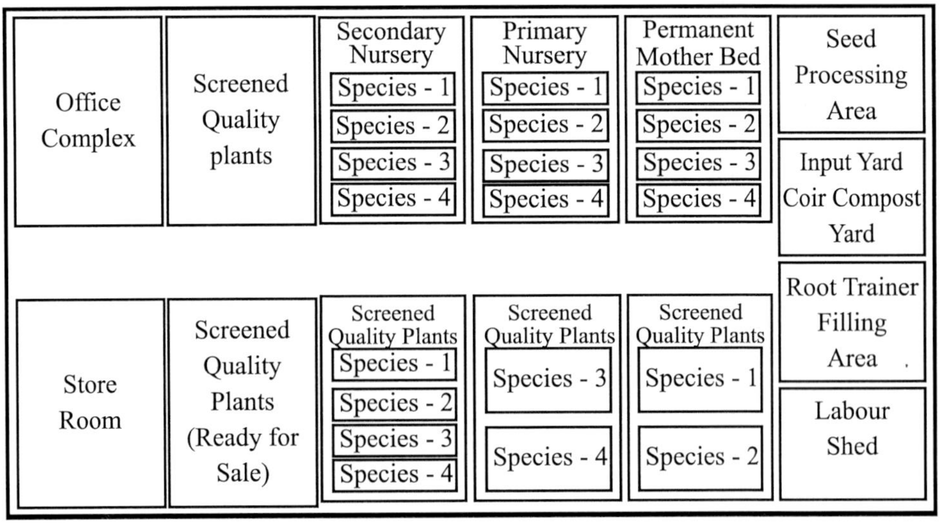

Fig. 1. Layout of a seedling nursery

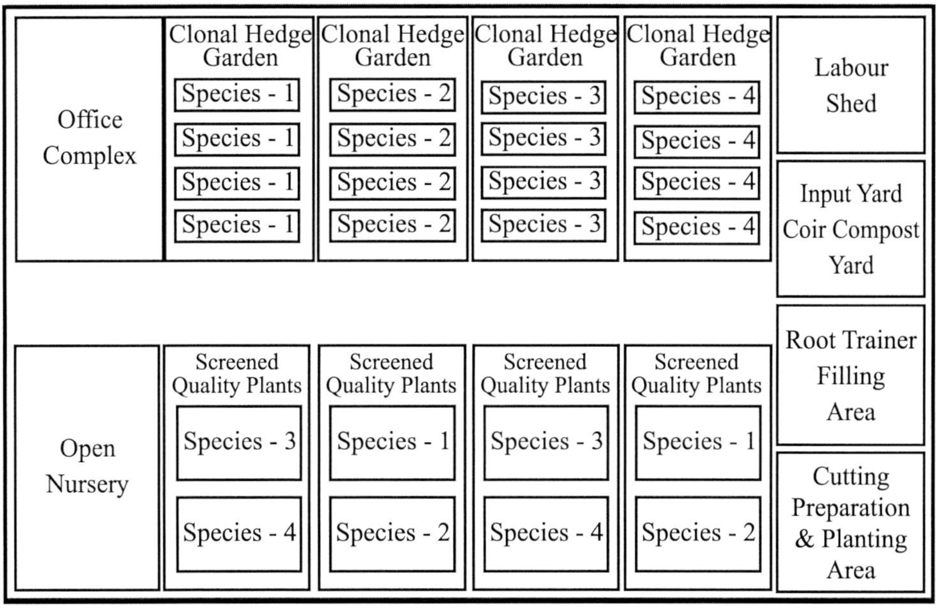

Fig. 2. Layout of a clonal nursery

Essential Requirements of a Centralized Nursery

1. **Protection fence:** The nursery area should be protected from the entry of cattle and intruders and also from the heavy wind. For this, a chain link fence and a row of trees all around the nursery should be established.
2. **Office:** An office should be established for administrative purpose besides maintaining the records and stock registers.
3. **Store:** A store room for storing nursery tools, chemicals, fertilizers, polythene bags etc. is very essential for a centralized nursery.
4. **Storage Yard:** Nursery potting mixture like compost, top soil, sand etc. need storage facility and hence a nursery storage yard is very essential.
5. **Water source:** To cater the water requirement of the nursery, sufficient bore wells, water tank (both underground and overhead) are required.
6. **Clonal hedge garden:** This is essential for maintenance of clones and for collection of cuttings at regular intervals.
7. **Mist chamber:** For mass multiplication and efficient rooting of clones, an organized mist chamber is essential.
8. **Hardening chamber:** The propagated plants from mist chamber need proper hardening and hence a hardening chamber of required size and capacity needs to be established.

9. **Working space:** Sufficient working space is required for preparing cuttings and grafting before they are placed inside the mist chamber.
10. **Waste pit and composite yard:** To dump unwanted materials and to prepare compost for use in nursery.
11. **Motorable road:** For transporting nursery materials like soil, sand, FYM etc. for raising nursery and for easy lifting of seedlings from the nursery to the planting site.
12. **Entry gate:** The entrance should be provided with a lockable iron gate.

Seed

The fertilized ripened ovule of a flowering plant containing an embryo and capable normally of germination to produce a new plant is called Seed.

Seed is a vital reproductive material used in all quality planting stock production. Quality seeds should be collected from seed orchards or Seed Production Areas (SPAs) established for the purpose of quality seed production and supply. The quantity and quality of the tree seeds produced by any species will ultimately determine its regenerating capacity.

Based on storage and germination capacity, the seed are classified into 4 types

1. Orthodox Seeds
2. Sub-Orthodox Seeds
3. Temperate-Recalcitrant Seeds
4. Recalcitrant Seeds

Orthodox Seeds

Seeds, which can withstand drying down to low moisture content of around 5% to 10% and successfully stored at low or sub freezing temperature for long periods.

e.g.: Acacia, Anthocephalus, Eucalyptus, Betula, Duabanga, Fraxinus, Pinus and *Picea* etc.

Sub-Orthodox Seeds

These seeds loose viability within a few months in open air. These can be stored under same condition as true orthodox, but only for six months to a maximum of six years in some cases, loss of viability ranges from 0% to 34% when stored at -5°C to -20°C and moisture content between 5 to 10 per cent.

e.g.: Abies, Juglans, Salix and Poplar.

Temperate-Recalcitrant Seeds

Seeds are desiccation sensitive and can be dried to 35 to 50 per cent moisture content of fresh weight. Storage temperature varies from 3°C to-3°C

e.g. : Acer, Aesculus and *Quercus.*

Recalcitrant Seeds

Seeds which cannot tolerate drying below a relatively high moisture content (often in the range of 20% to 50% net basis) and which cannot be stored successfully for long periods.

e.g.: Hollong, Mekai and other dipterocarps, cane, champ, neem, rubber, and members of family Lauraceae, etc.

These seeds are sensitive to low temperature, chilling damage and death may occur if stored in low temperature. These are most difficult group to store even for short period.

Based on the problems posed by the tree seeds at the time of sowing, they are broadly classified into two main types:

(a) Non-problem seeds

(b) Problem seeds

(a) Non-problem seeds

These are seeds which pose little / no problem in germination and have no dormancy period or do not require any pretreatment before sowing.

Example: *Eucalyptus, Casuarina*

(b) Problem seeds

Seeds are said to be problem seeds when they are hindered in the process of germination by various factors. The problems are mainly due to hand seed coat, hand should fatty oils, and chemical.

(i) Hard seed coat

Seeds of certain tree species have a hard seed coat which disallows water and air to penetrate the seed coat (testa) and break the physiological dormancy and promote germination.

Example: *Acacias, Leucaena, Cassia, Albizia, Delonix* etc.

(ii) Fibrous seed coat or hard shell

Some tree seeds have a hard and fibrous seed coat or shell which is very much impermeable to water and air and hence remain dormant till favourable conditions (natural scarification) occur.

Example: *Terminalia*, Teak, *Gmelina arborea* etc.

(iii) Seeds containing fatty oils

Certain tree species possess oil in their seeds and when exposed to room temperature lose their viability quickly.

Example: Neem, Pungam, *Bassia*, Kapok etc.

(iv) Chemical hindrance

Chemicals produced in the seeds of some tree species reduce their viability quickly after maturity.

Example: *Ailanthus* spp.

(v) Immature embryo

Seeds of certain trees will not mature even after the fruits are ripened. They require an after ripening period for the seeds to germinate.

Example: Pines etc.

Seed Treatment

The germination problems experienced by problem seeds can be overcome by the following seed treatment methods:

S.No.	Type of seed	Treatments
1.	Hard seed coat	(a) Soaking in hot water (b) Acid scarification using commercial grade conc. H_2SO_4
2.	Fibrous or hard shell	(a) Alternate soaking and shade drying
3.	Seeds with fatty oils	(a) Sowing immediately after collection and extraction of seed from fruit.
4.	Chemical hindrance	(a) Collect from the tree before fully nature and soaking in cold water for a day or two and then take up sowing.
5.	Immature embryo	(a) Cold or hot stratification.

Hot Water Soaking Method

Take three times of water when compared with the volume of seeds in a container and boil it. Remove the container from the oven and put the seeds into boiling water. Keep the seeds as such overnight and use them for sowing the next day morning. Care should be taken to see that the seeds are not dropped into the container when the water is boiling and do not exert pressure on the seed coat while sowing for it will lead to the expulsion of the cotyledons.

Acid Scarification

For acid scarification, commercial grade concentrated H_2SO_4 is used. Depending upon the seeds, the quantity of acid and duration of scarification will vary.

Example

Species	Concentration	Duration
Prosopis	150 ml / Kg	3 minutes
Acacias	200 ml / Kg	20-25 min
Cassias	200 ml / Kg	30-35 min
Delonix	250 ml /Kg	40-45 min

Seeds are placed in glass or plastic containers and the measured quantity of acid is added to the seeds and the seeds are stirred for a specific period. The seeds are then washed 8-10 times in cold water to remove the acid and then used for sowing.

Alternate Soaking and Shade Drying

In some tree species, fruits are used as sowing material instead of seeds. Example: Teak, *Terminalia* etc.

To facilitate easy germination, the fruits are initially soaked in cold water for 24 hours and then shade dried for 24 hr and this process is continued for 5-7 times before the fruits are sown. This process will cause swelling and shrinking of fruits which will lead to cracks on the hard shell and will allow water and air to enter the fruit and hence facilitate quick germination.

Other Methods of Pretreatment

(i) For pest and disease control

The seeds are dipped or coated or fumigated with suitable insecticide to avoid insect damage after sowing.

(ii) Seed hardening

After initial scarification, the seeds are soaked in 5% salt solution followed by drying and used for direct seeding which may facilitate for drought hardiness.

(iii) Seed Pellatization

Seeds are pelletized with fertilizers and insecticides using a filler and gum and used for aerial seeding.

(iv) Treatment with microbial cultures

To produce robust and healthy seedlings, seeds are sometimes treated with Rhizobial cultures, shade dried and used for sowing. This will help in fixing atmospheric nitrogen.

Phospho-bacterial cultures, *Frankia*, Ecto and Endomycorrhiza are some of the beneficial microbial organisms which could produce healthy seedlings. Seed treatment with microbial cultures should be provided after seed scarification, if any.

General Techniques Adopted for Tree Nursery

(1) Land leveling and site preparation

The area chosen for nursery should be thoroughly ploughed and hoed to a depth of 45 cm and completely removed off its unwanted materials. The clods are then broken and the area is levelled. Depending upon the type of mother bed (earthen beds or filled in polythene containers), the area is laid out. Tree seeds when sown directly into polybag containers and raised which is called direct sown nurseries or initially raised on mother beds and then transplanted into polybag containers which are called as transplanted nurseries. The beds are laid out in rectangular shape with their length east to west to avoid shade affect. The preparation of the mother bed and the transplanting bed are discussed further.

(2) Enrichment of the site

Well ploughed field is nourished with organic fertilizers like compost, manure or well decomposed farm yard manure (FYM) followed by formation of raised bed or sunken beds. Usually a height or depth of about 15 cm is given for the raised and sunken beds respectively and for species like Teak and *Terminalia,* the bed height is increased to 45 cm. To prevent sliding down of the bed, the side of the bed is consolidated by ramming. In places experiencing heavy rainfall, materials like wooden planks, bricks or coconut palm plates may be used. The upper portion of the mother bed (2.5-5 cm) is prepared with a soil composition of Red earth, sand and FYM sieved with a 2 mm sieve and mixed in the ratio of 3:1:1. In case of certain species like *Casuarina*, bamboo etc., the sand fraction should be provided through drainage channels around the nursery to prevent water stagnation. The top portion of the bed is properly levelled and now the bed is ready for sowing. In many cases, temporary shade is provided at a height of 3 feet to protect the germinating seedlings from ill effects of scorching sunlight and heavy rain.

(3) Soil mixture

The soil selected for potting mixture should be of less clay content so as to encourage easy root development and good drainage. The best potting mixture is red earth or native soil with less clay content, sand and FYM in the ratio of 3:1:1. In places where tank silt is available in plenty, one part could also be added. While filling with the above mixture, care should be taken to see that they are free from clods and they are thoroughly mixed.

(4) Containers

Size of the containers used will depend on the species and the purposes for which they are raised and economy during transportation. Most of the nurseries predominantly use polybags containers for seedling production and root trainers for clonal multiplication.

Sizes

- Polybags: 10 × 20 cm, 15 × 25 cm, 30 × 45 cm
- Root trainers: 90 cc, 150 cc, 300 cc and 600 cc

(5) Transplanting or primary nursery

Transplanting nursery is the nursery area wherein the seedlings are pricked out and planted from the mother bed or where seeds are sown directly and seedlings are raised before they are planted out. In this nursery, nursery seedlings from the mother bed are pricked out and planted in the poly containers filled with standard nursery mixture to develop complete plantlets for planting out.

(6) Size of the nursery

For economic production of tree seedlings, the size of the nursery should be sufficient enough to meet the seedling demands. A mother bed of 1 metre square is sufficient to produce 2000-2500 plantable Eucalyptus seedlings and transplanting bed of size 10 m × 1.2 m can hold 1700-2000 seedlings depending upon the filling of polybags of size 20 cm × 10 cm. Hence, based on the seedling requirment, the nursery area should be decided.

(7) Time of sowing

Every nursery man should carefully plan when he should sow the particular species so that seedlings of optimum age are readily available for planting at the beginning of the monsoon. For most of the species, nurseries are sown during February-March

e.g.: *Acacia, Eucalyptus, Casuarina, Leucaena* etc.

For species like Teak, it is taken up during September-October. Depending upon the optimum age required at the time of planting, the time of sowing should be adjusted.

(8) Sowing

Based on the size of the seeds and germination percentage, seeds are sown directly into the transplanting bed from where they are out planted or sown in mother bed and then transplanted into the transplanting bed. For example seeds of *Eucalyptus*, *Casuarina* etc. are very small in size or light and difficult to dibble into polybag containers and hence they are raised in mother beds and then transplanted into polybag containers. Though certain tree seeds have high germination percentage (80%), due to improper collection or storage, the germination percentage is usually low and if these seeds are directly dibbled into polythene bags, resowing has to be done in those empty bags which is not only a costly affair but also will lead to different aged seedlings in the nursery. In this case, the seeds may be prior tested for its germination percentage and the type of nursery technology to be adopted may then be decided. Seeds of tree species should be pure, graded and tested for its germination percentage before they are sown in the nurseries. The most popular methods of sowing are discussed below.

Methods of Sowing

(i) Broadcasting

In this method, seeds are spread uniformly over the mother bed and covered to thickness of the smallest diameter of the seed. This is usually taken up for small seeds (*Eucalyptus*), light seeds (*Casuarina*) and seeds with poor germination percentage.

(ii) Line sowing

Line sowing is adopted to prevent overcrowding of seedlings due to improper spreading during broadcasting and also to produce healthy seedling through proper spacing of seeds in the mother bed. This is also used in places wherein seedlings are directly taken out from the mother bed to the main field. In this method, lines are drawn on the mother bed and seeds are sown in these lines and covered. The distance between lines depend upon whether the seedlings stock used for transplanting into the transplanting bed are directly planted into the main field and also based on the germination capacity of the seeds. For seedlings that are to be transplanted, the line distance is close when compared to those which is taken directly for out planting from the mother bed. Line sowing is usually recommended for the species which has high germination percentage.

(iii) Dibbling

This method of sowing is adopted for bigger sized seeds with high germination percentage (80%). In case of seeds with low germination, seeds are dibbled with increased number per drill in which case the quantum of seed requirement will be high. In this method, seeds are sown into the containers by making a hole to a depth of one centimetre. Seeds dibbled at greater depth will have delayed germination or in some cases it may be killed due to lack of water or air.

Tips for Sowing

(a) Sowing should be taken up early in the morning or late evening.

(b) Beds and containers should be watered and soaked thoroughly before sowing.

(c) Sow the seeds shallow rather than deep and cover the seeds to the thickness of the smallest diameter of the seeds.

(d) Treat the seeds with some fungicides like Agrason or Bavistin to prevent fungal disease and ant removal.

(e) Scarified seeds should be given a coating of microbial cultures like *Rhizobium* to produce healthy seedlings.

(f) Species like *Terminalia*s with the fruits partially exposed to the sunlight will give early germination.

(g) When pre-sprouted seeds are used (will prevent gap after sowing due to non germinated seeds), care should be taken to see that the sprouts are not damaged while sowing.

(h) For those tree species which are amenable for transplanting, at the time of direct sowing into polybag containers a portion of the seeds may be sown in the mother bed of sand media and after 15 days, the germinated bags could be pricked with the seedlings from the mother bed. This will help in preventing different aged seedlings in the nursery and will not require grading at the time of planting.

(9) Seed requirement

The requirement of seeds of a particular species should be worked well in advance and procured or collected at the time of fruiting and kept ready for sowing. Never go in search of seeds after preparing the nursery bed. The quantity of seed required to plant one hectare is worked out using the following formula-

$$\text{Seed requirement} = \frac{\text{Ha} \times \text{N}}{\text{S} \times \text{PS}}$$

Ha = Size of annual planting programme for each species.

N = Intended number of seedlings to be planted per hectare.

S = An estimate of survival after planting and the need to gap fill.

PS = Number of plantable plants/kg of seeds.

(10) Watering

Beds should be kept constantly moist at the time of seed germination. During the early stages, the beds should be given a fine spray of water using rosecan. In case of fine seeds which are thinly covered, straw or twigs should be placed over the bed before watering. Watering should be done daily twice till one month after germination and then reduced. Depending upon the species and area for which they are raised, watering should be regulated to harden the seedlings in nursery. Watering is done at intervals of 1-7 days and subsequent watering is usually done by noticing the seedlings at ready to wilt stage. In large scale nurseries, sprinkler irrigation may reduce the cost of watering when put up on a permanent basis. After plants are grown up, watering can be done by percolation or hand splash or by rosecan.

(11) Weeding

For healthy growth, the seedling should be free from weeds which may affect the plant growth if they are not removed at the right time. Proper knowledge to distinguish weeds from the tree seedlings is a must at the early stages of weeding. Based on the intensity of weeds, the time of weeding should be decided by the nursery men.

(12) Temporary shade houses

To protect the top surface of the soil from drying up and protect the young seedling from scorching sunlight and dispersal by sudden down pour, temporary shade houses are erected over the mother beds. Sometimes, these shade houses can be erected permanently to cater the protection needs of germinating seedlings.

(13) Manuring

The size of the container and the nutrients in the soil will not be adequate to nourish the seedling for a long period. After three months of growth in the containers, these seedlings may be given irrigation with cow dung solution. For this, one part of fresh cow dung is mixed with 4 parts of water and kept in a container for 2 days. On the third day, the supernatant solutions is taken and to this, add 10 times of water and use this for irrigating the seedlings once in 15-20 days before shifting. Seed treatment with rhizobial cultures, VAM, Phosphabacteria, *Frankia* and spraying of Urea 1% or DAP application @ 2.5 gm/ polybag containers is also given to boost up the seedling growth. It is better to prefer organic form of fertilizer which will nourish the seedlings slowly and steadily and will help it to survive on waste lands which are characterized by poor nutrition.

(14) Shifting and root-pruning

Shifting of containers from its original place is a must to prevent the root from striking into the soil and also to break the top clogged soil caused by continuous irrigation which will prevent the entry of water and air. Fifteen – twenty days after germination or planting, check one or two seedlings at random in the bed and see whether they have struck roots ground and if so, do the shifting. After the first shifting operation, shifting of seedlings should be done once in 15-20 days interval. Weeding also should be combined with the shifting operation to minimize the cost of seedling production. Protruding roots out of the container should be pruned at the time of shifting to keep the roots compact and prevent them from penetrating into the soil.

(15) Thinning and Culling

In case of overcrowding of seedlings due to sowing of more number of seeds in the polybags or sowing or high germination percentage seeds on mother beds, thinning is advocated to produce healthy seedlings. Thinning should be done once the germination has been completed. While thinning care should be taken to retain those seedlings which have germinated earlier and are healthy and attention should be directed towards removal of late germinated and unhealthy seedlings. To increase production at the time of last shifting, weak, pest and disease attacked seedlings are culled out to retain only healthy plants in the nursery for out planting. The culling percentage can go up to 30%.

(16) Pest and disease control

To seeds are treated with quinylphos dust to protect them from ants. Young seedlings are often attacked by leaf eating caterpillar and leaf webbers for which any systemic insecticides can be used.

Common diseases observed in the nursery are damping off and root rot. Avoiding overcrowding of seedlings and allowing sunlight to reach the bed can control such diseases. At times when damping off is observed, avoid irrigation and apply Bavistin or Blue Copper @ 2 gm/lit of water and drench the bed. Pre treating seeds with agrason, thiram, captan will help to control the disease.

(17) Pricking

Seedling from the mother bed is pricked out into the transplanting bed from where they are planted out. The following points should be kept in mind at the time of pricking out of seedlings.

i. Prepare and keep ready the transplanting bed by watering it the previous day of pricking and erecting a shade net over it.

ii. Water the mother bed the previous day.

iii. Seedling shall be pulled out form the mother bed early in the morning or late evening and planted in the transplanting bed.

iv. The seedlings should be pricked without folding the roots and it should be well compacted.

v. The root portion of the pulled out seedling should be kept inside water in a container till they are planted.

vi. Water the seedlings daily twice, in the morning and in the evening.

The age of pricking will vary from species to species and this should be strictly followed or else the seedling may not survive after pricking.

Eg. *Eucalyptus, Casuarina, Ailanthus* - 30-40 days after sowing
Acacia nilotica, A.leucophloea etc. - 15-20 days after germination.
Bamboo - 40-60 days after sowing.

(18) Type of planting stock

The following types of planting stocks can be produced in the nursery.

(i) Bare rooted seedlings or naked

Bare rooted seedlings or naked seedlings are used for planting in sandy loam soils where water table is high or areas with high rainfall or with irrigation facilities. Here the seedlings are transplanted from the original mother bed to the transplanting bed at wider spacing from where they are pulled out and planted in the main field. Care should be taken not to injure the roots or keep the plants in the open so that the roots get dried up at the time of planting.

Example: Casuarina, Terminalias, Bamboos etc.

(ii) Polythene Containerized seedling

This is a common method wherein seedlings are raised is some containers for the prescribed period and then planted out.

Example: Eucalyptus, Casuarina, Acacia, Albizia etc.

(iii) Stumps

In this method, the seeds are sown on the mother bed and seedlings are raised. After the prescribed period, these seedlings are pulled out and the shoot portion (2.5 cm) root portion (22.5 cm) and side roots are pruned off and the left out material called the stump is used for planting.

Example: Teak

(iv) Root trainer seedlings

Seedlings are also raised in various types of root trainers and are used in commercial agroforestry plantation establishment.

Example: Casuarina, Eucalyptus etc.

(19) Nursery period

The age of seedlings at the time of planting will vary from species to species and also with the moisture status. In case of high rainfall areas or irrigated areas, seedlings of the age of 3-4 months old in the nursery could be planted out. In drier tracts, it is better to plant older seedlings of more than 6 months old. The common nursery period for effective planting stock is as follows.

Eg.	Eucalyptus, Casuarina, Leucaena	-	3-4 Months
	Acacias, *Ailanthus*, Neem etc.	-	6 months
	Bamboo	-	9-12 months
	Teak	-	10-12 months
	Tamarind	-	18 months

Nursery machinery, tools and materials

For effective functioning of a nursery, it should have the following machinery, tools and materials always available and in a good condition. The efficiency of the labourers are very well determined with the condition of the nursery tools and the knowledge of the work.

(i) Nursery machinery

Tractor with trailor, Tanker, Plough

This is required in all centralized nurseries to transport materials required for the nursery, ploughing and bush clearing operation. In case of failure of pump set in drawing water, the tractor with tanker can be used for transporting water and giving protective irrigation to the seedling.

Trowel Trolleys etc.

To carry small quantity of soil and other nursery materials into the nursery.

(ii) Nursery Tools

Mummutty, Spade, Shovel, Digging fork, Rake, Hand fork, Watering cans, Spraying pumps, Pen knife, Dudding knife, Secateurs, Measuring tape, Ranging poles, Stakes, Budding tapes, Dibbler (pointed stick) crowbar, Pickaxe etc.

(iii) Nursery Materials

Polybags, growth regulators (IAA, IBA), Fertile top soil, sand, FYM, seedlings brought out from the mist chamber before they are kept in the open

Nursery records

For effective management of nurseries proper records have to be maintained for planning, execution and monitoring the cost of production of seedlings. The following register should be maintained in all nurseries.

(i) Planning register

In this register, the species to be grown for the season, the number, how to be grown, and the input requirements etc should be forecasted and got in advance are entered.

(ii) Seed register

The seed materials to be purchased, their source, date of its collection, date of purchase, method of its storage etc have to be recorded to monitor the quality of the seeds. The germination percentage at the time of procurement should also be recorded for each species in this register. The date of issue of the seeds for sowing, germination per cent in the field should also be recorded.

(iii) Operation register

This register will have all details from sowing of a species till the seedlings are removed from the nursery for out planting.

(iv) Input register

In this all details on the cost of seeds, soil, sand, FYM, labor employed, water charges, Fertilizer cost etc. is to be recorded.

(v) Forecast register

Daily work or weakly work in a nursery has to be forecasted and done accordingly is recorded in this register.

Cost of seedling production

Any nursery could be a failure, if the cost of various inputs applied for seedling production is not calculated and planting stock produced accordingly. The cost of seedling production depends on the following factors.

(i) Location of the nursery

(ii) Availability of Sand, Soil, FYM, Water, Seeds etc.

(iii) Labour availability cost and efficiency

(iv) Technical knowhow on the production of seedlings.

Depending upon these factors the production, cost will vary from place to place which has to be monitored and worked out by the man in-charge of the nursery.

SILVICUTURAL PRACTICES

The success of any plantation development depends on its establishment and the associated management practices. Several plantations established in the country exhibited wide range of success and failures. The reason for failures are numerous, but the key reason is non adoption of required silvicultural practices specific to species. The following are the major silvicultural practices which decides the success of any plantation.

1. Choice of species
2. Selection of site
3. Planting
4. Spacing
5. Aftercare
6. Tending
7. Weeding
8. Cleaning
9. Thinning
10. Improvement felling
11. Girdling
12. Pruning
13. Climber control

1. Choice of species

- The tree species are selected according to the human needs such as protection, production, aesthetic and other environmental consideration.
- The species should have good timber, economic and other useful values.
- The species should be adaptable to the climatic conditions prevailing in the area.
- The soil of the region is considered while selecting a species for regeneration.
- The species should improve the soil fertility and reduce the soil erosion.
- It should be resistant to pest, disease and other adverse factors.
- It should be easily established with least cost.
- It should be fast growing one.
- It should consider the prevailing succession stages.
- It should not produce any allelopathic effect on other plants.
- The species should not alter the floristic composition of the forest.
- It should not become weed in future.

2. Selection of site

Site is very essential factor which decides the successful establishment of species. Testing of site for its quality and based on the quality species can be selected. Site selection is also based on the need and objective of plantation establishment. If exotic species is to be introduced, before commercial cultivation, systematic cross site examination of the species is very essential. Hence the site and its quality is very essential factor for implementing successful silvicultural practices.

3. Planting

Planting refers to the act or operation of setting the seedlings in the ground for propagation of plants.

Planting time

Climate plays a deciding role in determining the appropriate planting time. Newly planted trees do best when exposed to moderate temperature and rainfall. Seedlings need time to root and acclimatize before the onset of intense heat and dryness of summer or the freezing temperatures of winter. Hence, early spring or onset of monsoon is the best time for planting. In case of localities having severe winter, it is better to plant before the snow fall. This also varies with location to location and species to species. Planting may be of any one of the following kinds:

1. Monsoon planting
2. Pre-monsoon planting
3. Winter planting
4. Spring planting

Planting methods

The most common mistake when planting a tree is digging of a hole, which is both too deep and too narrow. In deep pits, roots don't have access to sufficient oxygen to ensure proper growth. In case of narrow pits, root structure can't expand sufficiently to nourish and properly anchor the tree. As a general rule, trees should be transplanted no deeper than the soil in which they were originally grown. The width of the hole should be at least 3 times the diameter of the root ball or container. This facilitates root establishment. Planning is done either by *entire planting or stump planting*.

i. Planting an entire plant

In this method, the whole plant is lifted from the nursery bed and planted in the field for further growing. This is done in any one of the following methods:

a. ***Planting naked root seedlings:*** In this method, the seedling is safely removed from the nursery bed and soil around the root is removed. Thereafter the roots are treated with biofertilizers and or fungicides before planting. These seedlings are planted in the centre of the pits prepared in the field. While planting, the roots of the seedlings are spread out and soil is pressed compactly so as to get a good contact. The collar region of the seedling should be 2-5 cm above the ground level. This method of planting could not yield better results for many species and hence not widely used.

b. ***Planting seedlings with ball of earth:*** Ball planting is defined as the planting of seedlings with a ball of earth containing their roots without any disturbance to roots. For this purpose, the mature seedlings are detached from the wetted nursery bed with a ball of earth of sufficient size. This is planted in a similar way as described above in the naked seedling planting. The main disadvantage with this method is the damage of balls during transportation. This method is somewhat labour intensive process.

c. ***Planting seedlings in containers:*** This is the safest method of planting among all kinds of planting. In this method of planting, the plant containers are removed before planting in the pits without disturbing the roots.

ii. Stump planting

Stump is a specially prepared seedling in which shoots are cut above the collar region and roots are pruned leaving primary roots at defined length. Stump planting refers to the planting of stumps in a field.

Stump preparation

Seedlings of 1-2 cm diameter or pencil size collar thickness are suitable for stump preparation. The seedlings should have only one tap root system with 30 cm or more length. Seedlings with two or more tap roots are discarded. The shoots of selected seedlings are cut at 2 cm above the collar region and removed. Similarly all the lateral and secondary roots are removed by a sharp cut. Thereafter the tap root is cut at a length of 22 cm from the collar region. This stump can be treated with biofertilizers inoculants and or pesticides for better protection. These stumps are ready for planting.

Advantages of stump planting

i. It is easier and cheaper one

ii. Transport of planting stock is simple

iii. It requires less labour

iv. Soil preparation in the field is easier

v. Seedling establishment and growth is faster

vi. Root development is good and deep

Planting Program involves

a. ***Boundary demarcation:*** The first things to do are to select the site and demarcate its boundary keeping the plot rectangular in shape.

b. ***Marking and felling:*** After demarcation, a stock map of the area is made and all trees leaving a few standards are marked for felling.

c. ***Soil and planting map preparation:*** After clearance the area is surveyed to prepare a soil map. During this survey the method of soil working is also decided

d. ***Fencing:*** The area of the plantation is fenced either by a cattle proof or any other kind depending on the surrounding environment.

e. ***Soil preparation:*** First the planting site is tractor ploughed. In case of burning, the roots have been removed. Thereafter soil preparation may be done either in the form of pits, or trenches.

f. ***Sowing and planting:*** Sowing should be neither very thick nor very thin. The nursery stock may be planted in proper season.

The following patterns of planting are followed in Indian Forestry:

- Line Planting
- Square planting
- Triangular planting
- Quincunx planting

a. Line Planting

In line planting, plants are planted at some spacing in lines which are also some distance apart.

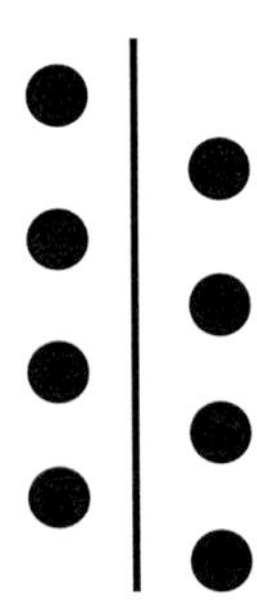

Line Planting

Calculate the Number of Plants Required

$$\text{Number of plants required} = \frac{100 \times 100}{\text{Distance of plants in lines} \times \text{Distance between the lines}}$$

b. Square planting

Square planting is defined as planting in square pattern, i.e., with plants occupying the four corners of each successive square. This is achieved by planting plants in lines at the same spacing as that of the line themselves.

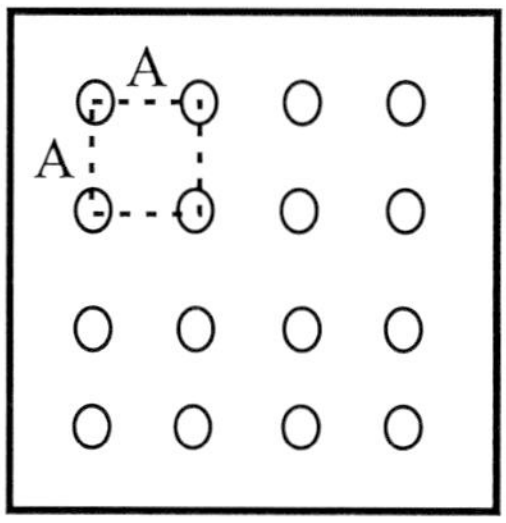

Square planting

$$\text{Number of plants required} = \frac{100 \times 100}{\text{Square of the planting distance}}$$

c. Triangular planting

Triangular planting is defined as the planting in the pattern of equilateral triangles, *i.e.,* with plants occupying the three corners of each adjacent equilateral triangle.

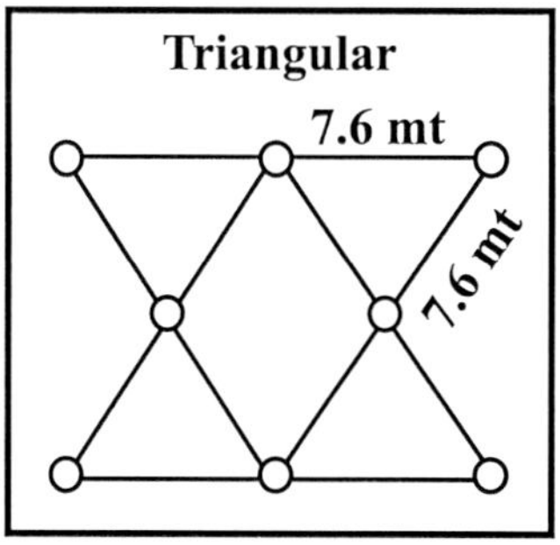

Triangular planting

$$\text{Number of plants required} = \frac{100 \times 100 \times 1.155}{\text{Square of the planting distance}}$$

(*i.e.* the side of the try angle)

d. Quincunx planting

Quincunx is defined as 'a group of five points, four of which from the corners of a square with the fifth as center'. Thus quincunx planting is that method of planting 'in which an extra plant is placed in the center of the each square of four plants'.

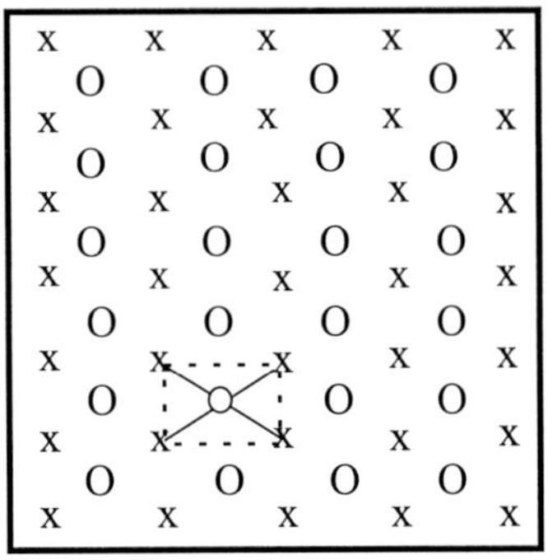

Quincunx planting

$$\text{Number of plant required} = \frac{2 \times 100 \times 100}{\text{Square of the side of the planting squre}}$$

4. Spacing

Spacing is defined as the distance between the plants put out in a plantation or standing crop. This is also referred as *espacement.* Based on distance between plants, spacing is classified as closer spacing and wider spacing.

Advantages of wider spacing

- It saves seed and plants
- The cost of plantation is reduced
- Labour requirement is less

Disadvantages of wider spacing

- There is a danger of soil deterioration as canopy takes longer time to close.
- Trees become branchy and produce knotty timber.
- The wood strength is poor.
- It results in higher weed growth.
- The weeding process is labour intensive and cost consuming.

Advantages of closer spacing

- It checks the weed growth.
- It reduces detoriation of soil.
- The thinning becomes easier and produces short term return.
- The boles of trees are cleaner due to natural pruning.

Disadvantages of close spacing

- It requires large quantities of seeds or plants.
- This increases the cost of plantation.
- The labour requirement is higher.

Factors influencing the spacing of plants

1. ***Rate of growth of species:*** Fast growing species needs close spacing and *vice versa.*
2. ***Habit of branching:*** The species characteristics of having large branches and crown requires wide spacing whereas conifer needs closer spacing.
3. ***Number of annual rings per centimeter:*** Closer spacing is preferable if more number of annual rings per centimeter is required.
4. ***Height of planting material:*** If the seedling height is higher, wider spacing is adopted. This is because of the rapid growth of the seedlings which occupies the space shortly and reduces weed problems.
5. ***Site factor:*** At better site conditions, close spacing is adopted as the site supports more trees per unit area.
6. ***Inter cultivation:*** In the case of taungya (raising agricultural crops in between trees in forest areas), the spacing between the trees should be higher than the normal. This is due to spacing for arable crop cultivation
7. ***Market for small sized timbers:*** In areas requiring demand for small woods, it is better to adopt closer spacing and *vice versa.*
8. ***Objectives of the production:*** If the objective of production is timber, wider spacing is advisable whereas for fuel wood production, closer spacing is recommended.
9. ***Cost:*** Based on the cost availability, spacing may be closer or wider. Closer spacing requires more cost for seedling requirement and *vice versa.*

Spacing of Some Important Species

Teak	1.8 m × 1.8 m to 2.6 m × 2.6
Eucalyptus	2 m × 2 m / 3 × 14 / 3 × 2 m/ 3 × 3 m
Casuarina	1 m × 1 m to 2 m × 2 m
Melia	3 m × 3 m to 10 m × 10 m
Dalbergia sissoo	2 m × 2 m to 5 m × 5 m
Gmelina arborea	4 m × 4 m to 6 m × 6 m
Subabul	1.5 m × 1.5 m to 3 m × 3 m
Ailanthus	4 m × 4 m to 6 m × 6 m
Silk cotton	5 m × 5 m to 8 m × 8 m

5. Aftercare

i. ***Beating up and weeding:*** Sowing and plantings are never cent percent successful and, therefore, casuality replacement either by sowing (or), replanting has to be carried out. Similarly weeding is a regular process to avoid competition.

ii. ***Fire and general protection:*** After weedings, the plantations require protection against grazing and fire. Properly fenced areas are safe as far as grazing is concerned but grazing guards have to be appointed to keep fence in proper state of repairs. With the onset of summer, the plantation has to be protected against fires by keeping watch guards.

6. Tending

In a forest, competition among plants can be increased with increasing age. In order to obtain a maximum yield the competition among the plants can be optimally managed by reducing the number of stems per unit area with the progress of age. All these can be covered under the tending operations.

Tending is defined as an operation carried out for the benefit of a forest crop at any stage of its life between the seedling and the mature stages.

This includes weeding, cleaning, thinning, improvement felling, pruning, climber cutting and girdling. This does not include regeneration felling, soil working, drainage, irrigation and controlled burning. Tending helps in producing higher quality timber and thus maximizes income. Tending differs from the cultural operation as indicated in next page.

Tending	Cultural operations
Tending is carried out for the benefit of forest crop by creating better environment	It is done for assisting regeneration, promote development and to decrease felling damage
This is not aims at natural regeneration	This aims at natural regeneration
It includes pruning but not controlled burning	It includes controlled burning but not pruning
Tending carried out from seedling stage	Cultural operations carried out after felling
It is carried out in both natural as well as artificial regeneration	This is carried out in silvicultural system relying primarily on natural regeneration

Limitations in Tending

- It has to be done for several times from seedling to maturity
- It requires considerable funds and staff
- This creates lobour shortage problem which in turn creates supervising problem

7. Weeding

Weeding may be defined as a tending done in the seedlings stage in nursery or in a forest crop. AJ involves the removal or cutting back of all weeds.

Objects of weeding

- To reduce root competition for nutrients
- To reduce transpirational water loss
- To facilitate cultural operations
- To improve light conditions to main crop
- To reduce the competition for space.

Weeding process

- Weeding must be done in plantations during the rains and stopped by the end of September
- It should be done before the suppression of seedlings by the weeds
- It is carried out for 3 years from planting
- Fast growth species requires less weeding than the slow growing species

8. Cleaning

Cleaning is defined as a tending operation done in a sapling crop which involves removal of inferior growth including individuals of favoured species. This is normally carried out when the inferior crop interfere with the main crop.

Objects of cleaning

- To improve light conditions
- To reduce root competition
- To reduce transpirational water loss

Operations in cleaning

- Clearing of shrubs and herbaceous growth
- Removal of individuals of inferior species
- Cutting back of the malformed or diseased individuals of the desired species
- Cutting and removal of climbers
- Singling of coppice shoots of the favoured species

9. Thinning

Thinning is a felling made in an immature stand for the purpose of improving the growth and form of the trees that remain, without permanently breaking the canopy. Thinning is a tending operation carried out in a crop beyond the sapling stage and up to the beginning of regeneration period.

Objects of Thinning

- Thinning helps to improve the hygiene of forest crop, by removing dead, dying and diseased trees.
- Trees which have lost in the struggle of life are usually weak and very much susceptible to be attacked by diseases and insects. Such trees are likely to be a source of infection for those which are fit in the struggle. Thinning helps to remove such trees which are infected with pest and diseases.
- Thinning helps to improve the growth and development of the retained trees due to availability of more space resulting in less competition and more availability of natural resources.
- Thinning helps to improve the horizontal growth of tree species thereby helps to achieve the diameter increment.

- Thinning helps to develop uniform and regularly managed crop suitable for varied domestic and industrial utility.
- Thinning helps to bring interim benefits through the harvest and sale of thinned materials.

Methods of Thinning

The following methods of thinning are normally practiced in all forestry activities

i. Mechanical thinning
ii. Ordinary thinning
iii. Crown thinning
iv. Free thinning
v. Craib's advance thinning
vi. Numerical thinning

I. Mechanical thinning

This type of thinning is usually applied in young crops or young plantations before the crown differentiation has taken place. There are two general systems followed in this type of thinning *viz.*

i. **Row-thinning :** Thinning in which trees are cut in line or rows throughout the area and

ii. **Spacing thinning:** Thinning in which trees at fixed intervals of distance are selected tor retention and all other are cut. The mechanical thinning since applied in the younger crops, it is not possible to identify the vigorous trees because most of them grow with more or less same vigour at an early age.

Example: In teak plantations of Kerala, first and second thinnings carried out at the age of 4 and 8 years are usually mechanical thinning and consists of the removal of alternate diagonal lines or rows of trees reducing the stocking from 2,500 trees per ha to 1250 trees per ha, after the first mechanical thinning. Then the population is reduced to 625 trees per hectare after second thinning with spacing increased from 2 m × 2 m to about 2.85 m × 2.85 m and then ultimately to 4m × 4m after second thinning. The thinning practices followed are shown in the Fig. 1.

Practice of mechanical thinning even in plantations becomes difficult in areas where casualties are more and already gaps exist in the area. The application of row thinning or space thinning is very much limited and depends upon the extent of natural regeneration, as species with dense natural regeneration can be thinned by either of these methods in earlier stages.

Some thumb rule methods have been developed to facilitate in execution by less technical staff. These are based on the relation between the average diameter of trees and the average spacing between the trees.

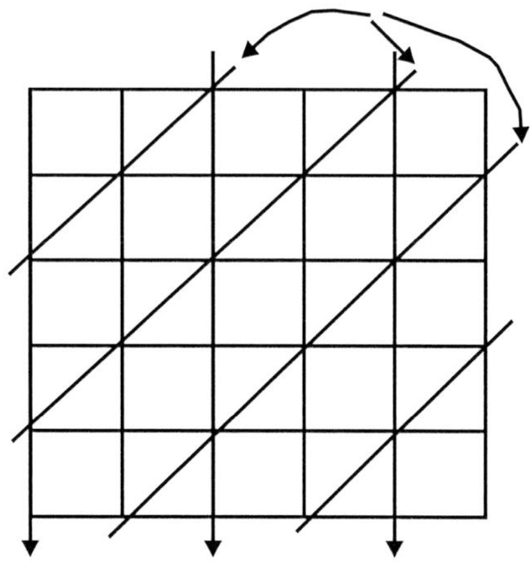

Fig 1. Spacing after the first and second mechanical thinning

The relationship between these two is usually a linear relation and can be represented by

$$d = aD + b$$

Where

a and b are the two constants,

d the spacing in feet

D the average diameter of trees in inches.

Disadvantages of Mechanical thinning

i. The main disadvantage of this method is arbitrary selection of the trees in rows or lines which may result in removal of good trees and leaving of poor ones in a crop.

ii. The method is good for areas with relatively uniform productivity alone.

iii. However, if the productivity or area is varying, it may result in the removal of the trees which may be growing on better place and are likely to produce more yield. Removal of wolf trees from natural stands or plantations along with thinning is called wolfing.

II. Ordinary thinning

This is also called as 'low thinning' or 'German thinning' or 'thinning from the below'. In ordinary thinning, the trees are removed from lower crown classes. The trees which are removed in this thinning belong to diseased trees of all classes, suppressed, dominated and some dominants depending upon the grades of the thinning. Ordinary thinning has been devised to follow the nature i.e. those trees, which have been unsuccessful in the struggle of existence are removed first.

Advantages and applications of ordinary thinning

i. It is most suited for light demander species, Eg: chir, teak, sal, sissoo, semal, gamhari, etc. because in light demander species, the suppressed and dominated individuals do not usually retain power of recuperation and are not likely to respond to improved conditions of growth due to thinning,

ii. This method of thinning is useful and may be economically applied, in species and areas where small size timber has a market,

iii. The method is simple in execution and only ordinary skill is required for selection of trees, therefore, where thinning is to be executed by less trained persons, it is better to adopt this thinning,

iv. Removal of lower crown classes includes several diseases and insect infected individuals. Thus, this thinning practice improves the hygiene of the crop.

v. The method is suitable for areas where there is no danger of soil deterioration as a result of removal of smaller sizes trees.

vi. The method is useful in forests infested with climbers and where there is risk of crown fire.

vii. In this type of thinning, trees of lower crown classes are removed to accelerate the growth of remaining trees. It is reported in case of several species that growth in dominant trees is stimulated only when some gap is created and root competition is considerably reduced by removing some trees in the top canopy and removal of smaller trees does not significantly affect the growth. Only heavier grades of thinning are likely to produce desired results. Therefore, this method of thinning with proper grade has to be selected taking into consideration the growth, soil loss, etc.

viii. Cutting of lower crown classes at later stages of development helps to obtain natural regeneration.

Disadvantages of ordinary thinning

i. This method of thinning rests on removal of trees of lower crown classes. In several areas, troublesome and desirable trees and shrubs in the understorey may develop. In sal, it is reported that dense understorey develops in areas where heavy opening are created in lower crowns.

ii. This thinning is designed to be executed when trees have been in competition for sufficiently a long time to develop crown differentiation. This means that whole crop including the dominants has already suffered due to adverse effect of competition,

iii. Sometimes, lower crown classes may be useful particularly when predominants and dominants are not able to utilize the site perfectly. The surplus of the nutrients store is utilized by lower crown classes and is kept in cycling.

iv. If there is no demand of small trees, it is often difficult to utilize the thinned material economically. When thinned material remains undisposed, it increases fire hazard and does not improve the hygiene of the crop

v. The removal of trees of lower crown classes may expose the soil. This increases soil erosion particularly in slopy areas.

Grades of ordinary thinning

The following grades of ordinary thinning are practiced.

S.No	Different grades	Characteristics and intensity of thinning	Significance of thinning
1	**Light thinning (A grade)**	Involves removal of dead, dying, diseased arid suppressed trees i.e. classes V, IV and III	No practical use Useful for comparative research on the effect of thinning on the increment, volume, etc.
2	**Moderate thinning (B grade)**	Removal of defective dominated stems, whips, Branchy advance growth i.e. classes V, IV, III, II(b), I (d) and an occasional I (c).	Only of little use in ordinary practice because of reduced influence on the remaining stems
3	**Heavy thinning (C grade)**	Removal of dominated stems and defective co- dominants without making lasting gaps in the canopy i.e. classes V, IV, III, II and I (b), I (c) and I (d).	standard grade in vogue for yield table computation
4	**Very heavy thinning (D grade)**	This grade differs from the above in the way that that it also takes some of the dominants without breaking the canopy. i.e. class V, IV, III, II, 1(b), 1(c), 1(d) and some 1(a).	Improves the quality of trees retained by improving crown expansion and cause considerable girth increment.
5	**Extremely heavy thinning (E grade)**	The heaviest thinning that can be done in a crop without making permanent gaps in the canopy removal of even class 1(a)	Useful for research purpose

III. Crown thinning

Crown thinning is a method of thinning which is primarily directed to the dominant trees in a regular crop, the less promising ones being removed in the interest of the best individuals; the dominated and suppressed stems are retained unless they are dead, drying or diseased. This method of thinning is known as French thinning because of its origin in France. This is also called thinning from the above or high thinning to differentiate it from ordinary thinning.

The beneficial effects of removal of lower crown classes on the growth of trees of top crown classes is doubtful and some of the trees of lower crown classes need to be retained for protection of the site and checking invasion of undesired under storey. This method of thinning therefore, aims at removing the inferior trees from among the dominant class. The bulk of trees of lower crown classes, e.g. dominated and suppressed, are ignored and allowed to grow after each thinning. In this method of thinning, trees from upper crown classes are removed in order to favour the growth and development of most promising frees of the same class. The selection of trees to be retained after thinning depends primarily on the health, vigour, form and species. If all the trees are of good health, vigour, form and species, codominants are removed which are interfering with the growth of predominant trees. If there is a choice between relatively poor predominant and promising codominant, the codominants are favoured for retention. In crown thinning, the object is to promote the growth of trees which show signs of success in the competition and promise a good value as a future crop.

Crown thinning is suited for a number of conditions,

i. It is best suited for moderately shade tolerant species, e.g. sal, deodar, etc. because, in such species, dominated and suppressed trees retain the power of recuperation for a long period,

ii. It is suited for areas where there is no demand for small sized timer

iii. It is suitable for forests where considerable damage due to frost, snow, wind, etc. occur. The vacancy created by the death of dominant trees due to these factors is filled by the trees of lower crown classes

iv. The method is especially suited for area where there is danger of .site deterioration due to removal of trees of lower crown classes.

Advantages of crown thinning

i. Crown thinning provides better environment for growth and development of retained dominant trees, they grow faster than in ordinary thinning. The production of timber is of big size and of better quality.

ii. The trees of lower crown classes are not removed under this method of thinning. They keep weeds and shrub growth under control,

iii. The pressure of trees of lower crown classes results better pruning of side branches. This helps in production of better timber,

iv. This method of thinning helps in protection of the site and reduces the damages due to frost snow, wind, etc.

Disadvantages of Crown thinning

i. There is higher root competition for moisture and nutrients adversely affecting the growth of dominants when growth factors are limiting,

ii. Crown thinning presents problems and obstruction in felling, logging and extraction of thinned material,

iii. Crown thinning is more flexible method than ordinary thinning. It requires greater skill in execution. It requires more strict numerical check on the trees to be removed,

iv. The trees of lower crown classes, if, diseased or insect infested, serve as a nucleus of infection in the retained crop. Therefore, closer look on suppressed and dominated trees would be necessary.

Grades of crown thinning

The following two grades of crown thinning are recognized.

S.No	Different grades	Characteristics and intensity of thinning	Significance of thinning
1	Light crown thinning (L.C. grade):	Removal of dead, dying and diseased trees	Favours development of the best available trees.
2	Heavy crown thinning (H.C. grade)	Removal of crown classes V, IV, 1(d), 1(c), most of 1(b) some of 1(a)	even distribution of trees over the area

IV. Free thinning

Free thinning is a modification of crown thinning and is also called 'single stem silviculture', or 'elite thinning".

In this method, 'elites' are first selected in number appropriate to the size or age of the crop with special regard to their stem form and to their even spacing. In this method, instead of treating the crop as a whole, attention is concentrated for the provision of optimum growing space for future stems throughout, the growing period of the crop.

The required number of future stems corresponding to the rotation age are first selected and retained more or less uniformly distributed all over the area. In selection of future stems, co-dominants or even dominated trees of perfect stem form and standing in the appropriate position relative to neighbouring elites stems, may be preferred to quite fair but less acceptable dominants. The elites may be marked for easy identification. Once the elites are identified, the balance crop is considered from the point of view of their effect on the elites. If there removal is likely to help the elites, they are cut otherwise they are retained for the protection of the site.

V. Advance thinning

In this method, thinning is carried out before the competition among individual trees has set in; the competition is never allowed to take place. This may cause loss in the total volume production, but ensures maximum growth of the elite trees. This method was suggested by Craib (1939) for plantations in South Africa and usually known as Craib advance thinning.

In the earlier stages of development, there is an intense competition among the plants for water, nutrients, light and space. This competition is regulated by removing the surplus individuals and diverting the resources to the residual trees. Sometimes the trees which are freed take 3 to 4 years for starting faster rate of growth. This leads to the conclusion that greatest increment is possible only when no suppression has yet set in the crop.

VI. Numerical thinning

Various factors such as number of stems per ha and diameter at breast height relationship, N/H^2 (number of stems per ha and height relationship), tree area ratio, bole area, basal area, etc., can be very good guide for the optimum density of the crop at a particular age and site and can be used as a numerical check on thinning,

(i) **N/D relationship:** The relationship between number of stems per ha and diameter for several Indian timber species has not been found independent of site quality.

Stand density index is the number of trees corresponding to a given standard average diameter. Such numerical parameter on the basis of relationships between number of trees and average diameter are suggested on the basis of standard relationship of number of stems per hectare and average diameter of the crop.

(ii) **N/H^2 relationship:** For a given site quality, the optimum number of trees per unit area is found to be inversely proportional to the square of the height of the stand i.e. N/H^2. However this method is not used in Indian forests.

10. Improvement felling

Improvement felling refers to the removal of less valuable trees in a crop in the interest of valuable individuals. It is applicable to mixed as well as irregular crops. In this following operations are carried out:

- Removal of dead, dying and diseased individuals
- Felling of saleable unsound and over mature trees
- Felling of unsound or badly shaped mature and immature trees
- Thinning of congested groups of poles
- Cutting back of damaged seedlings and saplings (v/') Removal of undesirable under growth
- Climber cuttings

11. Girdling

Girdling is defined as cutting through bark and other living layers of woods in a continuous incision all around the pole of a tree. This is the economical method of killing of inferior trees inside the forests. Girdling of stem restricts flow of photosynthates from leaves to roots. Thus girdling causes death of roots by starvation. This is restricted to certain species only. Girdling enhances more fire hazard inside the forests.

12. Pruning

Pruning refers to the removal of live or dead branches or multiple leaders from standing trees for the improvement of the trees or its timber. It may be of dry pruning (removal of dead branches) or green pruning (removal of live branches).

It is classified as natural pruning and artificial pruning. Natural pruning is defined as the natural death and fall of branches of standing trees. It is also referred as self pruning. This pruning is carried out by natural agencies such as wind, drought, ice, snow, etc.

Artificial pruning is defined as the pruning done by the forester without waiting for nature to do it. This reduces the cost of formation and rotation.

Bud pruning is a special kind of pruning which is defined as the rubbing off the lateral buds to prevent the development of branches as a measure to obtain knot-free timber. As it prevents development of branches from the lateral buds, it reduces the pruning at later stages.

13. Climber control

Climbers are usually cut regularly to avoid its damage to the growth and developed of infected trees. Thin climbers needs complete uprooting. In case of thick climbers it may be cut both at base and one meter above. Then climbers are removed and burned.

CHAPTER - 7

SILVICULTURE OF COMMERCIALLY IMPORTANT TREES

INTRODUCTION

1. SILVICULTURAL OF EUCALYPTUS

Species		:	*Eucalyptus camaldulensis,* *Eucalyptus tereticornis*
Family		:	Myrtaceae
Common Name		:	Red gum, Mysore gum
Locality Factors	**Altitude**	:	0 - 1000 m
	Mean Annual Rainfall	:	600 – 1500 mm
	Mean Annual Temperature	:	2°C to 32°C
	Soil type	:	• Sandy loams • Gravels and Alluvial Soil
Phenology	**Flowering**	:	Flowering occurs twice a year • May to June • October to November
	Fruit ripening	:	• July • December
Silvicultural Characters		:	• Strong light demander • Coppice - coppices freely and vigorously
Nursery Techniques	**Seed Propagation**	:	• Seed treatment – Nil • Seeds are raised in the mother bed

	Vegetative Propagation	: • Germinated seedlings are transplanted • Six month old seedlings are ready for planting • Clonal technology – mini clonal technology • Establishment of clonal mother garden • management of mother garden with irrigation and fertilizer • Induction of micro shoots • Cutting of 5 to 10 cm size • Planting in root trainer filled with coir compost • Root initiation within 21 days • Hardening 45 to 60 days • 90 days old plants are ready for planting
Silvicultural Treatment	**Spacing**	: 3m × 1.35 m / 3m × 1.5 m /3.0 × 2.0 m
	Pit Size	: 30 cm × 30 cm × 30 cm
	Basal Application	: 250g of Vermi-compost or 2 kg of Farmyard manure per pit with 50-100 g DAP
	Planting Time	: June to October
	Irrigation	: 3 to 6 litres per day
	Fertilizer Application	: 2 kg of FYM and 100g – 500g all 19 once in every six month
	Weeding	: Two weedings per annum
	Pruning	: Self Pruning
	Thinning	: Only dead and diseased
Rate of growth		: • Fast growing short rotation trees. • It yields an average of 125 - 150 tonnes / ha in 3 years under irrigated condition
Rotation		: • 3 Years for Biomass, • 5-6 Years for Pulpwood • 6-8 Years for Ply wood
Utilization		: • Eucalyptus is an excellent raw material for pulp and paper production due to higher pulp yield which ranged between 44 and 48 percent.

- It is a good plywood due to clearily and quality
- Fuel wood and charcoal due to high calorific value of over 4500 kcal / kg.
- Wood is strong and used in particle board and hardboard industries.
- The leaves of the Eucalyptus species are rich in essential oils

2. SILVICULTURAL OF CASUARINA

Species		:	*Casuarina equisetifolia,* *Casuarina junghunhniana*
Family		:	Casuarinaceae
Common Name		:	Beefwood, She-oak
Locality Factors	**Altitude**	:	0 - 1000 m
	Mean Annual Rainfall	:	900 – 3800 mm
	Mean Annual Temperature	:	10 - 47°C
	Soil type	:	• Best in loose, fine coastal sands. • For inland conditions - Well drained sandy loams soils. • It tolerates Lateritic and red soils and also saline, alkaline and acidic conditions.
Phenology	**Flowering**	:	Flowering occurs twice a year • February to April • September to October
	Fruit ripening	:	• June • December
Silvicultural Characters		:	• Strong light demander and drought resistant. • Susceptible to fire. • Coppices badly. • Tolerate waterlogged conditions
Nursery Techniques	**Seed Propagation**	:	• Seed treatment – Nil • Seeds are raised in the mother bed

		• Germinated seedlings are transplanted • Six month old seedlings are ready for planting
	Vegetative Propagation	: • Clonal technology – mini clonal technology • Establishment of clonal mother garden • management of mother garden with irrigation and fertilizer • Induction of micro shoots • Cutting of 5 to 10 cm size • Planting in root trainer filled with coir compost • Root initiation within 21 days • Hardening 45 to 60 days • 90 days old plants are ready for planting
Silvicultural Treatment	**Spacing**	: 1.5 x 1.5 m
	Pit Size	: 30 cm x 30 cm x 30 cm
	Basal Application	: 250g of Vermi-compost or 2kg of Farmyard manure per pit with 50 g DAP
	Planting Time	: South West and North East monsoon
	Irrigation	: 6 to 8 litters per day
	Fertilizer Application	: 100 kg urea first year in three dozes 150 kg DAP at four dozes in second year 150 kg complex in three dozes in third year
	Weeding	: Two weedings per annum
	Pruning	: Once in every six months
	Thinning	: Dead and diseases
Rate of growth		: • High Yielding short rotation trees. • It yields an average of 100 to 150 tonnes / ha in 3 years.
Rotation		: • Rotation age of Casuarina is 36 months
Utilization		: • Pulp wood: Pulp yield is more than 47%. • Fuel Wood: Calorific value is 4950 Cal/Kg. • Wind breaks • Poles

3. SILVICULTURAL OF MELIA

Species		:	*Melia dubia*
Family		:	Meliaceae
Common Name		:	Malabar Neem, Melia
Locality Factors	**Altitude**	:	1500 to 1800 m
	Mean Annual Rainfall	:	800 – 1000 mm
	Mean Annual Temperature	:	32 - 40°C
	Soil type	:	Deep, well drained sandy loam soils
Phenology	**Flowering**	:	November - January
	Fruit ripening	:	January – February
Silvicultural Characters		:	• Light demander • Susceptible to damage by fires • Saplings suffer from browsing
Nursery Techniques	**Seed Propagation**	:	• Seed has to extracted from this stony endocarp and treated with GA at 100 ppm for overnight. • Seed are sown on sand bed and germination starts in 30 days • Germination is only 40% • Germinated seedlings are transplanted in poly bags • Six month old seedlings area ready for planting
	Vegetative Propagation	:	• Clonal technology – mini clonal technology • Establishment of clonal mother garden • Management of mother garden with irrigation and fertilizer • Induction of micro shoots • Cutting of 5 to 10 cm size • Planting in root trainer filled with coir compost • Root initiation within 21 days • Hardening 45 to 60 days • 90 days old plants are ready for planting

	Spacing	:	• Plywood: 4m x 4m (or) 5m × 5m (or) 6 m × 6m • Pulp & Plywood: a) 6'× 6' (First two years)
	Pit Size	:	30cm × 30cm × 30cm
	Planting Time	:	June to October
	Irrigation	:	6-8 litre (day through drip irrigation)
Silvicultural Treatment	**Fertilizer Application**	:	Mixture of compost and organic fertilizers, bio-fertilizer and planting (25-50 g), respectively, to be applied to the pits. Implication 50 g DAP
	Weeding	:	Annual
	Pruning	:	Annual
	Thinning	:	• Thinning of alternate rows at the beginning of 3rd year • Thinning alternate diagonals at the beginning of 5th year • Final harvest at the beginning of 7th year
Rate of growth		:	• For pulpwood: 100-150 tonnes/ha in three years. • For Plywood: 200 tonnes/ha in six years
Rotation		:	• For Plywood: 5-7 Years • For Pulp and Paper: 24-36 months
Utilization		:	• It is a good plywood due to medium density end high veneer recoverly. • It is a good pulpwood due to increased poly yield. • The wood is used for packing cases, cigar boxes, ceiling planks, building purposes, agricultural implements, pencils, match boxes, splints and Kattamarams. • It is suitable for musical instruments, tea boxes and plyboard. • It is a good fuel wood (Calorific value: 3,400 - 4,100 cal.) • The fruit of the plant is bitter. It is considered anthelmintic.

4. SILVICULTURE OF TEAK

Species		:	*Tectona grandis*
Family		:	*Verbenaceae*
Common Name		:	Teak
Locality Factors	**Altitude**	:	0 to 1200 m
	Mean Annual Rainfall	:	1000 - 5000 mm
	Mean Annual Temperature	:	2°C to 48°C
	Soil type	:	Deep and well-drained soil Fertile Alluvial-colluvial soil
Phenology	**Flowering**	:	January to April
	Fruit ripening	:	May to July
Silvicultural Characters		:	• Strong light demander • Sensitive to frost and drought • Good coppicer and pollards vigorously
Nursery Techniques	**Seed Propagation**	:	• Seeds are treated with alternate wetting and drying for 14 days • Seeds sown in mother bed • Germinations stars after 3rd week • 9 – 12 months old seedlings are lifted for stump preparation • Stump size 2.5 cm shoot portion 22.5 root portion • Stump are used directly for planting
	Vegetative Propagation	:	It can also be multiplied throgh rooting of cutting.
Silvicultural Treatment	**Spacing**	:	2m x 2m
	Pit Size	:	30 cm x 30 cm x 30 cm
	Basal Application	:	2 kg of farmyard manure, 100 g DAP
	Planting Time	:	June – July or September - October

Irrigation	:	
Fertilizer Application	:	2kg of FYM, 100g of complex fertilizer and 100g – 300g all 19 once in every six month
Weeding	:	Annual
Pruning	:	Necessary, bivine a year
Thinning	:	• Thinning cycle of 4, 8, 12, 18, 26 and 36 years have been followed for 50 years rotation. • In Tamil Nadu the thinning cycle of 5, 10, 18, 25 and 36 years are followed for 50 years rotation. • In both the cases, the first two thinning are mechanical and the rest are Silvicultural at C grade thinning.
Rate of growth	:	Teak is a fast grower and attains a height of 15 feet to 20 feet in one year under well managed condition.
Rotation	:	• 20 years under well irrigated and managed condition • 40-60 years under forest site condition • 15 years under bund plantations.
Utilization	:	• Teak is a moderately strong timber with a density of 660 kg/m^3 and is preferred as a most suitable timber both for domestic and industrial utility. • Teak is known as a renowned timber due to durability, dimensional stability, working quality and resistant to termites. • Teak wood is used in all construction purpose such as beams, columns, doors, windows, flooring, paneling etc. • It is one of the best timbers for furniture and cabinet making wagon and railway cadges. • For marine construction and ship building teak is preferred due to dimensional stability.

5. SILVICULTURE OF AILANTHUS

Species		: *Ailanthus excelsa*
Family		: *Simaroubaceae*
Common Name		: Tree of heaven
Locality Factors	**Altitude**	: 0 to 900 m
	Mean Annual Rainfall	: 500 - 1900 mm
	Mean Annual Temperature	: 12.5°C to 47.5°C
	Soil type	: Porous sandy loams
Phenology	**Flowering**	: February to March
	Fruit ripening	: April to may
Silvicultural Characters		: • Strong light demander • Sensitive to drought • Moderately frost tender • Coppices well and produces root suckers freely. • Susceptible to water logging areas
Nursery Techniques	**Seed Propagation**	: • De winged Seeds soaked in cold water for 5- 7 days by replacing fresh water daily • The seeds are dibbled in poly bag and water daily • Six month old seedlings are ready for planting.
	Vegetative Propagation	: -
Silvicultural Treatment	**Spacing**	: 5m × 5m / 6m x 6m
	Pit Size	: 30 cm x 30 cm x 30 cm or 45 cm x 45 cm x 45 cm
	Basal Application	: 2 kg of farmyard manure, 100 g DAP
	Planting Time	: • The area is cleared and pits are dug out in the month of February - March and the soil is allowed to weather.

			• The planting in pits is carried out in the month of July.
	Irrigation	:	Dryland crop
	Fertilizer Application	:	5kg of FYM and 100g – 200g complex fertilizer yearly once in a year during monsoon season
	Weeding	:	Timely and regular weeding for the first two years are very essential
	Pruning	:	-
	Thinning	:	The first silvicultural thinning may be carried out in the third or fourth year when the tree attains a height of 10-12 m.
Rate of growth		:	Ailanthus is a slow growing trees and attains a height of 10 to 15 m. in six year under well managed condition.
Rotation		:	6 – 8 Years under irrigated condition and over 10 year in dryland
Utilization		:	• This species is extensively used for making matchwood boxes and match splints. • The wood is extensively used in cottage industries for making wooden toys and cheap quality cricket bats. • The tree is used for making packing cases and wooden boxes. • The leaves are rated as highly palatable and protein rich nutritious fodder for sheep and goats and are said to augment milk production. • It is suitable for sloppy, degraded and denuded areas and wasteland. • It is also yields gum of inferior quality.

6. SILVICULTURE OF SUBABUL

Species		:	*Leucaena leucocephala*
Family		:	Fabaceae
Common Name		:	Subabul
Locality	**Altitude**	:	0 to 1500 m

Factors	**Mean Annual Rainfall**	:	650 - 3000 mm
	Mean Annual Temperature	:	15°C to 36°C
	Soil type	:	• Calcareous soils • Saline soils and on alkaline soils up to pH 8
Phenology	**Flowering**	:	Two flowering season • July – November • February - May
	Fruit ripening	:	• December • June
Silvicultural Characters		:	• Strong light demander • Vigorous coppicer • Moderate frost tender • Drought resistance
Nursery Techniques	**Seed Propagation**	:	• Seeds are treated with concerated sulphuric acid and sown directlyin the poly bag • 3 – 4 months old seedlings are ready for planting
	Vegetative Propagation	:	
Silvicultural Treatment	**Spacing**	:	1.5m x 1.5m; 2m x 2m (for plywood); 3m x 3m (for timber)
	Pit Size	:	30 cm x 30 cm x 30 cm
	Basal Application	:	2 kg of farmyard manure, 50 g of super phosphate, 50 g DAP
	Planting Time	:	2-4 month seedlings can be used to planting out in the month of July.
	Irrigation	:	-
	Fertilizer Application	:	2 kg of FYM and 100g – 500g all 19 every six month once
	Weeding	:	3 month once for first two years
	Pruning	:	Regular during first one year
	Thinning	:	-

Rate of growth	:	• High yielding short rotation tree • It yield an average of 100 tonnes/ha in 3 to 4 years
Rotation	:	• 3-6 Years depends on location • It is an excellent plywood
Utilization	:	• Subabul has hard heavy wood (about 800 kg/m) and medium density wood. • Subabul is one the highest quality and most palatable fodder trees. • Subabul is an excellent firewood species with a specific gravity of 0.45-0.55 and a calorific value of 4600 cal/kg.

7. SILVICULTURAL OF SANDALWOOD

Species		:	*Santalum album*
Family		:	*Santalaceae*
Common Name		:	East Indian Sandalwood
Locality Factors	**Altitude**	:	90 – 1500 m
	Mean Annual Rainfall	:	500-2000 mm
	Mean Annual Temperature	:	15-35°C
	Soil type	:	• Sandy clayey red lateritic loamy even in black cotton soil • Red ferruginous (iron) loam over lying on metamorphic rocks • Rocky ground and stony or gravelly soils
Phenology	**Flowering**	:	Two Flowering Season • February – April • October - November
	Fruit ripening	:	• May • December
Silvicultural Characters		:	• Shade bearer • Root suckers freely

Nursery Techniques	**Seed Propagation**	:	• Coppices fairly well • The seeds exhibits initial dormancy for 3-4 weeks and after 4 weeks it starts germination which is about 60% • The uniform and very good germination can be obtained by soaking seeds with 0.05% gibberlic acid over night. • The sandal seedlings are transplanted along with host plants viz., *Casuarina, Cajanus cajan, Albizia, Alternenthra, Amaranthus* etc.
	Vegetative Propagation	:	Root cuttings of sandalwood gives only 20 % success.
Silvicultural Treatment	**Spacing**	:	3 × 3 m to 5 × 5 m
	Pit Size	:	30 × 30 × 30 cm (or) 40 × 40 × 40 cm (or) 60 × 60 × 60 cm
	Basal Application	:	Soil mixture with neem cake 25-50 g / pits, Chlorpyrifos 2 g powder / pit
	Planting Time	:	Monsoon Season
	Host	:	• Sandal has association with over 150 species of host. • Albizia, Terminalia, Lagerstroemia, Dalbergia, Casuarina, *Acacia nilotica*, *Pongamia pinnata, Wrightia tinctoria* and *Cassia siamea* are the major host plants.
	Irrigation	:	It is a dryland crop
	Fertilizer Application	:	During soil working periods application of farmyard manure @ 5 kg / plant
	Weeding	:	Yearly once
	Pruning	:	Pruning is essential to get good heartwood formation
	Thinning	:	-
Rate of growth		:	• Slow growing species • The heartwood formation in sandal starts after 10 years. • The heartwood forms at the rate of 1 kg/annum after 20 years.

Rotation	:	Physical rotation. The dead and naturally fallen trees are harvested
Utilization	:	• **Sapwood:** Sapwood is white and scentless used for manufacture of agarbattis. • **Heartwood:** Heartwood of sandal is moderately hard, heavy and strongly scented, wood and oil are used in incense, perfumes, soap making and medicines. • **Religious:** Sandal is considered sacred by Hindus. • **Essential oil:** Valuable oil, 'the sandal oil', is distilled from the heartwood (4-13%) and is used in perfumery, soap making and medicines. Seed oil: Seeds yield oil that can be used in the manufacture of paint. • **Medicinal uses:** The wood is bitter, dry, antipyretic, aphrodisiac useful in diseases of the heart, burning sensation, cold, bronchitis, vaginal discharges and small pox.

8. SILVICULTURAL OF REDSANDER

Species		:	*Pterocarpus santalinus*
Family		:	Leguminaceae
Common Name		:	Red sander, Red sandalwood
Locality Factors	**Altitude**	:	150 – 900 m
	Mean Annual Rainfall	:	350 -1350 mm
	Mean Annual Temperature	:	12 – 47°C
	Soil type	:	• Properly drained red soil • Dry rocky soil • Quartzite, shale, limestone and lateritic soil
Phenology	**Flowering**	:	April to June
	Fruit ripening	:	February to March
Silvicultural Characters		:	• Strong light demander • Resistance to fire

			• Excellent coppice • Drought resistance even at juvenile stage • Produce root suckers freely
Nursery Techniques	**Seed Propagation**	:	• The seeds are soaked with cowdung slurry for 72 hours or seeds are soaked with water for 72 hours with frequent change of water for every 12 hours. • Seed rate per bed: 1 kg • Germination Percent: 60-70%
	Vegetative Propagation	:	• Stump Planting • One year old seedlings are preferred for making stumps. • The stumps should contain 25-30 cm of roots and 10 – 15 cm of shoot. • Survival percentage is 87 %
Silvicultural Treatment	**Spacing**	:	3 m X 3 m to 6 m X 6 m
	Pit Size	:	30cm x 30cm x 30cm
	Basal Application	:	250g of Vermi-compost or 2kg of Farmyard manure per pit with 50-100 g DAP
	Planting Time	:	September to December Rainy season
	Irrigation	:	Regular irrigation is essential for initial three years of planting
	Fertilizer Application	:	Farmyard manure @ 2 kg / plant/ year DAP @ 100 g / Plant (Annual application)
	Weeding	:	2 weedings per year based on weed pressure.
	Pruning	:	Pruning is essential to obtain straight bole.
	Thinning	:	Thinning is practiced in 5 year cycle
Rate of growth		:	• Slow growing tree • The maximum height and girth were 12 m and 66 cm respectively • Wavy grained tree is preferable.
Rotation		:	25 years and above
Utilization		:	• Wood is fine red colour and beautifully streaked. Wood weight is 900 – 1265 mkg / cum with very strong and extremely hard.

- Timber: Excellent timber with little shrinkage.
- Musical instruments: wood is used as manufacturing of special musical instrument "Shamisen"
- The dye santalin extracted from wood is used as medicinal values.
- Leaves are good fodder.

9. SILVICULTURE OF KADAM

Species		:	*Neolamarckia cadamba*
Family		:	*Rubiaceae*
Common Name		:	Kadamba, Japon, Kalempayam, Vellaikkatambu
Locality Factors	**Altitude**	:	300-800 m
	Mean Annual Rainfall	:	300 - 1600 m
	Mean Annual Temperature	:	5 - 32°C
	Soil type	:	Prefers well drained entisols
Phenology	**Flowering**	:	May to June
	Fruit ripening	:	September to February
Silvicultural Characters		:	• Light demander • Tree coppices well
Nursery Techniques	**Seed Propagation**	:	The young seedlings are highly susceptible to weeds and should be weeded regularly. 2-month seedlings can be transplanted in nursery beds or into polythene bags, where they can be retained before planting at the start of the monsoon rains
	Vegetative Propagation	:	• Clonal technology – mini clonal • Establishment of clonal mother garden • management of mother garden with irrigation and fertilizer • Induction of micro shoots • Cutting of 5 to 10 cm

			• Planting in root trainer filled with coir compost with the treatment of IBA 1000 ppm • Root initiation within 30 days • Hardening 45 to 60 days • 4-6 months old plants are ready for planting
Silvicultural Treatment	**Spacing**	:	3m x 3 m
	Pit Size	:	30 cm x 30 cm x 30 cm
	Basal Application	:	250g of Vermi-compost or 2kg of Farmyard manure per pit with 50 g DAP
	Planting Time	:	June to October
	Irrigation	:	4 to 8 litters per day
	Fertilizer Application	:	2kg of FYM and 100g – 250g complex fertilizer once in every 3 months
	Weeding	:	Two weedings per annum
	Pruning	:	Twice per year
	Thinning	:	Only dead and diseased
Rate of growth		:	• Fast growing short rotation trees. • It yield an average of 125 - 150 tonnes / ha in 3 years.
Rotation		:	• 3-4 Years for Pulpwood • 6-8 Years for Ply wood
Utilization		:	• The wood has a density of 400-560 kg/cu m at 15% moisture content, a fine to medium texture; straight grain; low luster and has no characteristic odour or taste. It is easy to work with hand and machine tools, cuts cleanly, gives a very good surface and is easy to nail • The timber is used for plywood, light construction, pulp and paper, boxes and crates, dug-out canoes, and furniture components. Kadam yields a pulp of satisfactory brightness and performance as a handsheet.

- A yellow dye can be obtained from the root bark
- Kadam flowers are an important raw material in the production of 'attar', which are Indian perfumes with sandalwood (*Santalum* spp.) base in which one of the essences is absorbed through hydro-distillation
- The dried bark is used to relieve fever and as a tonic. An extract of the leaves serves as a mouth gargle

10. SILVICULTURAL OF POPLAR

Species		:	*Populus deltoides*
Family		:	*Salicaceae*
Common Name		:	Eastern Cotton wood, Poplar
Locality Factors	**Altitude**	:	300 – 1000 m
	Mean Annual Rainfall	:	600 – 1500 mm
	Mean Annual Temperature	:	4-35°C
	Soil type	:	• Fine, sandy loams, and fairly stiff clays • Moist, well-drained, fine, sandy loams or silts
Phenology	**Flowering**	:	April
	Fruit ripening	:	June
Silvicultural Characters		:	• Strong light demander • Fire tender • Does not produce root suckers
Nursery Techniques	**Seed Propagation**	:	Seeds are very minute and low germination percentage
	Vegetative Propagation	:	Conal Propagation Planting Entire Transplants (ETPs)
Silvicultural	**Spacing**	:	3 × 3 m to 5 × 5 m

Treatment	**Pit Size**	:	30 × 30 × 30 cm
	Basal Application	:	2 Kg of FYM / pits, 50 g of superphosphate and 50 gram of Urea
	Planting Time	:	May - October
	Irrigation	:	5 – 6 litres of water with 4 days interval.
	Fertilizer Application	:	• Farmyard manure @ 5 kg / plant • 100 – 200 g of DAP / Plant / Year
	Weeding	:	Yearly Twice
	Pruning	:	Pruning is essential to obtain straight stem
	Thinning	:	Selective and systematic thinning is needed

Rate of growth :
- Fast growing species
- 75 tonnes/ ha

Rotation : 4 – 12 years based on utility of wood

Utilization :
- **Timber:** The timber is used principally for lumber, veneer, pulpwood and excelsior.
- **Fuel:** Wood is used for fuel.
- **Fibre:** A good fibre and pulp is obtained
- **Fodder:** *P. deltoides* has been used extensively as a fodder species for sheep, goats and other livestock.
- **Medicinal:** The bark is used to treat rheumatism, gout and scurvy and infections of the chest, kidneys and stomach.

■■■■

CHAPTER – 8

FOREST MENSURATION

Measurement of trees is termed as Dendrometry or Forest Mensuration. It is the application of mathematical principles in estimating forest growth and development. The term Mensuration is derived from the latin word "**Mensura**" which means measurement. In forestry, measurements are very important in doing business transactions, assessing the growth potential and valuing the forest resources. Hence, Mensuration is a critical subject for all professional forestry graduates, wood and wood based industries and others involved in forest based business transactions. This chapter on Forest Mensuration incorporates the following sub divisions.

1. Definition
2. Objectives
3. Diameter Measurement
4. Height Measurement
5. Measurement of tree form
6. Volume estimation

DEFINITION

Forest mensuration is defined as that branch of forestry which deals with the determination of dimensions viz., diameter, height, volume, form, age and increment of single trees, stands or whole woods, either standing or after felling.

OBJECTIVES

- Measurement is a basis for sales and all business transactions.
- Measurement is a basis for management of forests leading to ensuring productivity on a sustainable basis.
- Measurement is essential for all research related activities leading to screening varieties for increased biomass attributes, resistant towards biotic and abiotic factors and augmenting the quality of wood.
- Measurement is essential for forecasting the feature demand based on the actual quantity of wood and wood resources available and hence supports the government is planning the wood availability.

- Measurement is essential for import and export of wood and wood resources based on the actual demand and supply within the country.

Measurement of trees

Trees are measured for the following factors viz., diameter or girth, height, form and volume of tree species in order to assess the quantity of wood and wood products. Before going for actual estimation the following factors are to be known.

(i) Accuracy

Measurements of trees are done based on the assumption that trees are cylindrical in nature and accordingly measurements are made and the volumes are calculated. But in reality no tree is true cylinder and hence assuming tree as a cylinder may create variability and the measurement may not give full accuracy rather it may lead to relative accuracy.

Relative accuracy means maximum possible accuracy that are obtained through measurement. Relative accuracy is due to the following characteristics.

1. Biological nature and characteristics of the trees
2. Varying degrees of growth and development
3. Varying methods of felling and the associated conversion process
4. Utility of different instruments for the same purpose
5. Personal variability

Units of Measurement

In India, during early forest management programme British system of measurement has been followed but all the measurement activities carried out today uses the metric system or French system or CGS system of measurement.

Parameter	Unit of measurement	
	British systems	**Metric systems**
Height	Feet without decimal	Meter with first decimal
Diameter	Inches with first decimal	Centimeter with odd decimal
Girth	Feet and inches	Meter and Centimeter
Area	Square feet	Square meter or centimeter
Volume	Cubic feet	Cubic meter

DIAMETER MEASUREMENTS

Measurement of diameter is very essential and it has been fixed that breast height as the standard height for measurement of diameter or girth of any standing trees.

Breast height is the universally accepted standard height for measuring girth, diameters and basal area of standing trees. The breast height followed in India is 1.37 m.

The diameter or girth measurements are referred to as D.B.H. (or d.b.h.) or G.B.H. (or g.b.h.) respectively. Unless otherwise stated, they are over bark (O.B. or o.b.) measurements and consequently, sometimes specified as D.B.H. (O.B.) or d.b.h. (o.b.) and G.B.H. (O.B.) or g.b.h. (o.b.) respectively. The diameter or girth measurement with bark is termed as dbh (ob) or gbh (ob) and the measurement after debarking is termed as dbh (ub) or gbh (ub).

Standard Rules for diameter at breast height measurement

The following are the standard rules governing breast height measurements:

1. The breast height should be marked at 1.37 m (4 ft 6 in) from the base and above the ground level and it can be marked by using a stick and by using the paint the position of breast height should be marked with a cross mark.

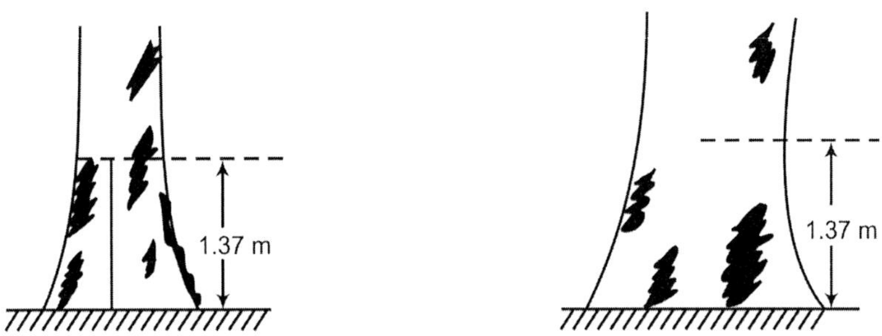

2. In case of trees growing on slopping ground the diameter at breast height should be measured on the uphill side only.
3. Suppose when the tree is leaning, the dbh should be measured along the tree stem and particularly on the leaning side, when it grows on flat ground. Suppose the leaning occurs in sloping area the dbh should be measured on the uphill side.
4. Suppose when the tree trunk is abnormal at 1.37 m, dbh should be measured either above or below the point of abnormality based on judgment by the observer and the measurements can be made.

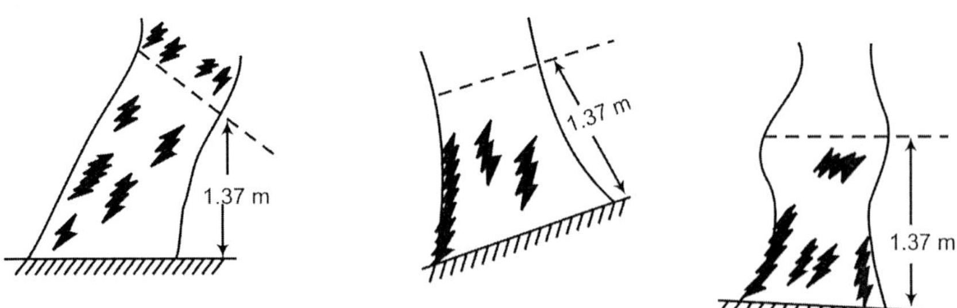

5. During the growth and development of trees forking or multiple branching occurs at several stages. Suppose when the forking occurs at 1.37 meters, based on the observers judgment the tree can be measured either singly or considered as double trees depending on the place of measurement.

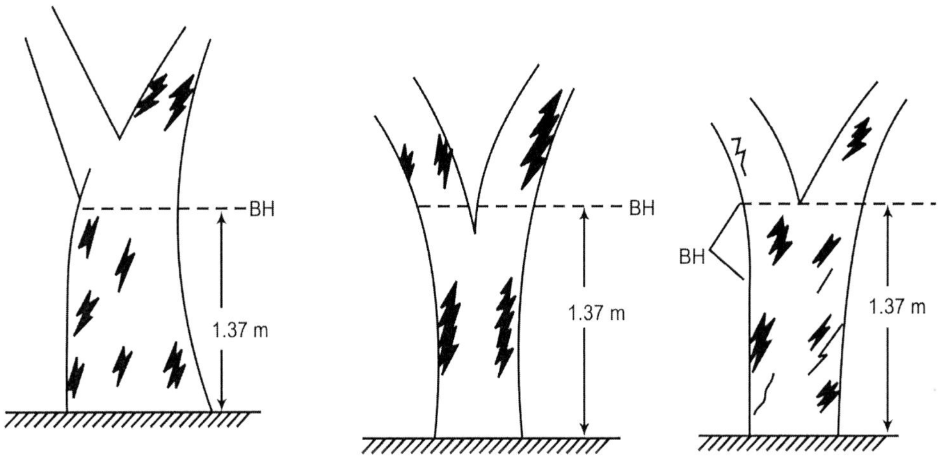

6. Many trees like Banyan and Terminalias exhibit buttress formation from the base and upto breast height level. In such cases, the diameter at breast height should be taken at the lowest point above which the abnormality is not witnessed.

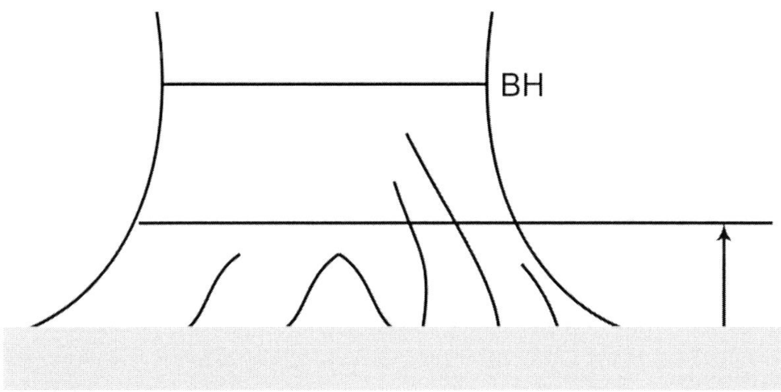

7. Before measurement the secondary growth on the stem of the trees like creepers, climbers, lichens, moss and lose bark must be removed.
8. Diameter at breast height should be measured in cm and to the nearest multiple of 2mm.

Instruments used in diameter measurement

In India, predominantly three instruments viz., Wooden scale, Callipers are Tape commonly used.

(i) Wooden scale

It is used by most carpenters for diameter estimation of felled trees and it is available in two ranges viz., 30 cm and 60 cm scale. It can be easily folded and carried over to any place of measurement very easily but applicable only for felled trees.

(ii) Tree Callipers

Tree Callipers are used for diameter measurement of standing trees. It consists of graduated rule with two arms. The left arm is fixed and the right arm is movable. The calipers of different sizes of 50 cm, 75cm, and 100 cm are available in India for use by researchers and professional foresters.

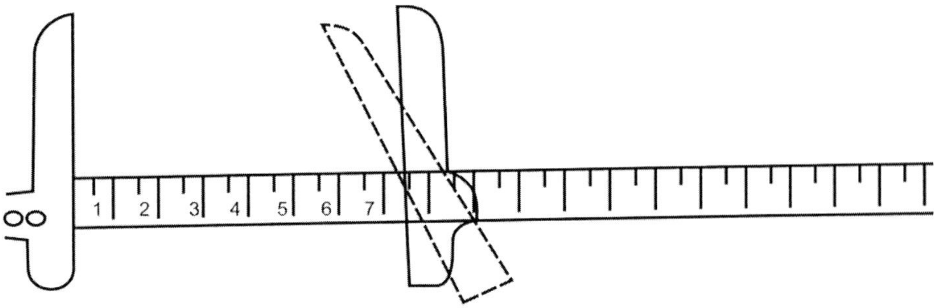

Advantages

- Tree calipers are useful in measuring diameters directly on the tree.
- It is amenable for easy to transport and operation by unskilled labour.
- The errors are both positive and negative and therefore they may neutralize the errors and ensure accurate results.

Disadvantages

- When the callipers are not in adjustment they may not give accurate results.

- The big sized callipers experienced difficult in carrying inside the forest area.
- The calliper measurement requires two measurements which may be difficult in difficult sites and need to either over estimation and under estimation of diameter measurement.

(iii) Tailor's tape

Tape is predominantly used to measure girth of trees and logs. The most commonly used tapes are 150 cm and it can be easily carried to any part of the forests and ensure easy and fast measurement. Taps are available in steel metallic and cloth but the cloth based taps are preferred by the users.

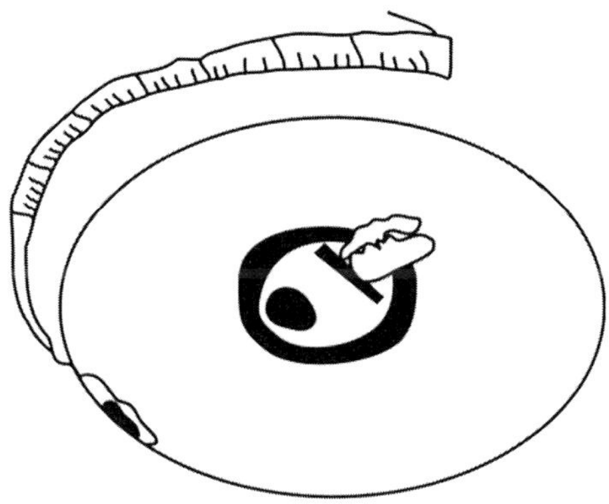

Advantages

- Tape is easy to carry and fast in measurement.
- Tape needs only one measurement and covers the entire circumference of the tree and ensures sustainability accuracy of measurements.
- The error in tape is always positive.

Disadvantages

- The girth measurement will be exaggerated if the tree has rough bark.
- Though tape requires only one measurement covering the full circumference of the tree the observer cannot see the full circumference and hence the breast height location may vary which lead to errors in measurement.
- The various quality tape lead to variability in measurement.

Diameter and girth classes

The diameter classes followed and used in India are 2 cm, 5 cm and 10 cm and the girth classes are 5 cm, 15 cm and 30 cm.

S.No.	Diameter classes	Intervals	Maturity
1.	2cm class	0-2, 2-4, 4-6, 6-8 etc.	Trees with less than 30 cm diameter
2.	5cm class	0-5, 5-10, 10-15, 15-20 etc.	Trees with 30-50 cm diameter
3.	10 cm class	0-10, 10-20, 20-30, 30-40 etc.	Trees over 50 cm diameter

HEIGHTS MEASUREMENTS

The second major measurement of tree is the tree height which decides the site quality of the locality. The height of the tree is the straight line distance from the tip of the leading shoot (or from the highest point of the crown where there is no leader) to the ground level, usually measured on slopes from the uphill side of the trees.

The trees are generally characterized based on its commercial utility and accordingly the height of tree is classified as follows.

1. **Bole height :** It is the height of the tree which varied between the ground level and crown point.
2. **Commercial bole height:** It is the height of the stem or bole that is usually fit for utilization as timber.
3. **Height of standard timber :** It is the height of the bole from the ground level upto the point where average diameter over bark is 20 cm.
4. **Stump height:** It is the height of the top of the stump above ground after felling.
5. **Crown length:** It is the vertical measurement of the crown of a tree from the tip to the point half way between the lowest green branches forming green crown all round and the lowest green branch on the bole.
6. **Crown height:** It is the height of the crown as measured vertically from the ground level to the point half way between the lowest green branch and the green branches forming green crown all round.

Methods of tree height measurement

Trees are measured continuously for estimating volume, vigour and growth till the inception of scientific forest management in the country. Traditionally tree heights are measured by the following three methods viz., Ocular, Non instrumental and Instrumental methods.

(a) Ocular estimation

The experienced foresters and the commercial timber merchants as on today uses ocular estimation as one of the standard method to estimate height of the tree. However it needs experience and inexperienced estimator may either under estimate or over estimate the height of the tree.

(b) Non-instrumental estimation

Tree heights are also estimated based on the shadow of the tree or the similar method. Accordingly the following two methods are practiced in early ages and the scientific instrumental invention in height measurement expelled the utility of these methods. However they are furnished only for academic interest.

(i) Shadow method

Here the shadow of tree is measured and the associated tree height is estimated deploying a pole. The tree to be measured is marked and a pole of known height is taken. Firstly the shadow of pole is measured and then the shadow of marked tree is measured. The actual tree height is then estimated using the following formula.

$$AB = \frac{BC \times DE}{EF}$$

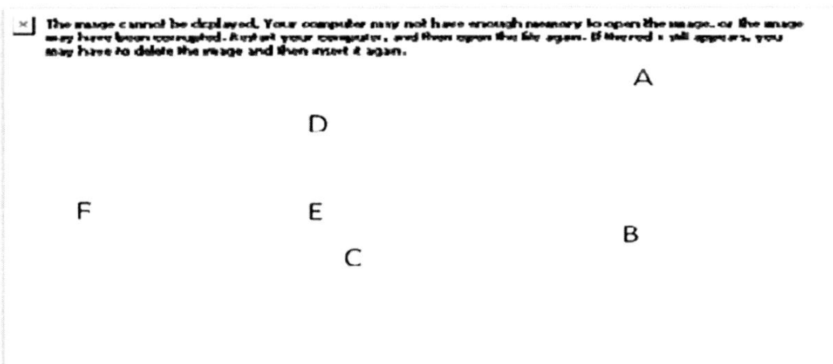

Where, AB is the tree height

DE is the pole height

BC is the length of the shadow of the tree and

EF is the shadow of the pole

(ii) Single pole method

1. Here, the estimator holds a pole of about 1.5 m height vertically at arm's length in one hand in such a way that the portion of the pole above the hand is equal in length to the distance of the pole from the eye.

2. Without changing the position of hand with reference to the eye, the observer moves slowly forward and backward till the line of sight to the tip of the tree passes through the tip of the pole and that to the base of the tree through the point where the pole is held by hand.
3. This means that the portion of the pole above the hand covers the tree completely.
4. The height of the tree is then equal to the distance of the observer's eye from the base of the tree as shown below:

Let *A B* be the tree and *De* pole about 1.5 m long, held at *b* vertically so that distance from observers eye F to *E* is equal to DE

Therefore, $AB = \frac{CB \times DE}{CE}$

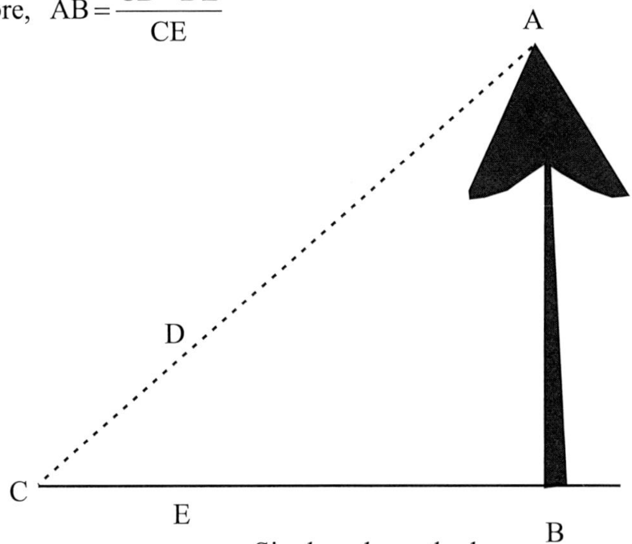

Single pole method

Since DE = CE, AB = CB

(c) Instrumental methods

The science and technology development has invented wide range of instruments to measure height of the tree. These instruments are designed based on the principle of similar triangle or trigonometric methods.

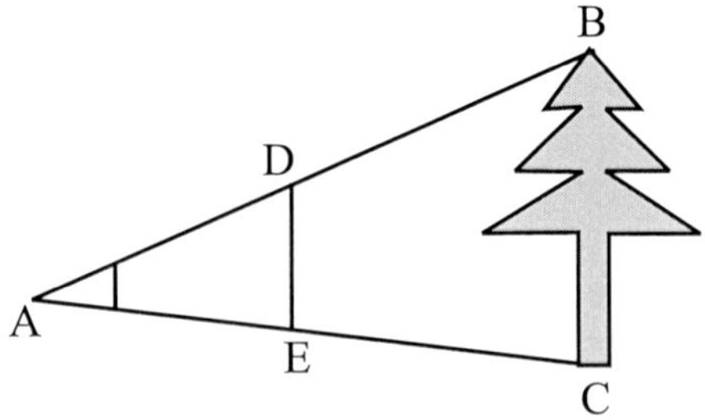

Principle of similar triangle

This principle has the following rules.

1. Each angle of a triangle is equal to its corresponding angle of the other triangle
2. Each side of a triangle is proportional to the corresponding side of the other triangle and
3. One angle of triangle is equal to one angle of the other and the corresponding sides which subtend the equal angles are proportional

ADE and ABC are similar triangles. Thus,

$$BC: DE = AC: AE$$

$$\text{Height (BC)} = \frac{DE \times AC}{AE}$$

S.No.	Principle	Instruments
1.	Trigonometric	1. Brandis Hypsometer 2. Abney's level 3. Altimeter 4. Relaskop
2.	Similar triangle	5. Christen's Hypsometer 6. Smythies Hypsometer

Trigonometric principles

Trigonometrical ratios are used for height estimation through the following method.

1. Tangent method
2. Sine method

1. Tangent method

Tangent method means that the height of the tree is calculated with the help of the tangents of the angles to the top and base of the tree and the distance of the observer from the tree. As the position of the observer with respect to the tree will differ according to the terrain, the procedures of calculating heights of trees have to be modified accordingly.

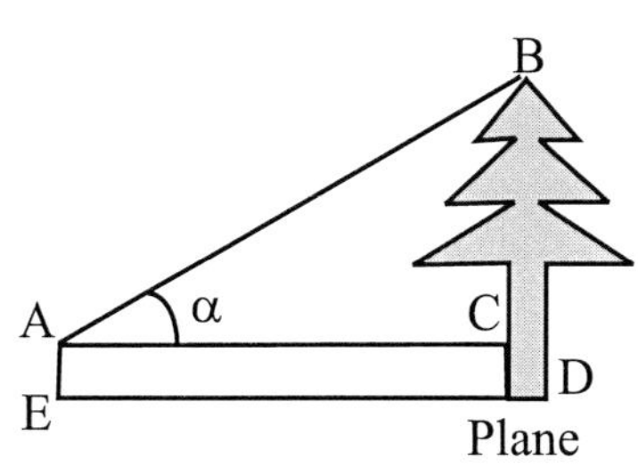

On level ground

The height of the tree is calculated with the help of the tangents of the angles to the top and base of the tree and the distance of the observer from the tree.

E.g.: BD = AE+ ED Tan α

Possible errors in height measurement

The following are the principal sources of error in height measurement:

- Instrumental errors
- Personal errors
- Errors due to measurement
- Errors due to observation
- Errors due to lean of trees
- Tree stem form

MEASUREMENT OF TREE FORM

Tree form

Tree form is the rate of tapper of a stem and the tapper is decrease in diameter from base of the tree upwards.

Principle of a tree form

The tree tapper or tree form has been explained by several authors and several theories have been putforth. However the most widely accepted theory is that of Metzger's theory. He considered tree stem as a cantilever beam of uniform size acting against the bending force of wind.

The wind pressure acts on the tree crown is transferred to the lower parts of the stem and the greatest pressure is exerted on the tree base hence the tree reinforces and resist the growing pressure.

Hence the growth material is accumulated more at the base to resist the incoming pressure and decreases towards the top. Trees growing in isolation have rapidly tapering stem to counteract the resistance and the trees growing in dense will have cylindrical poles.

Methods of studying form

The study of form is beset with difficulties because it cannot be fixed by one or two measurements. This is due to the fact that the form varies not only from tree to tree but also in the different parts of the same tree. Variation in bark thickness not only from stand to stand and tree to tree but also in the different parts of the same tree adds another dimension to the problem. However, the form of a tree can be studied in one of the following ways:

- By comparison of standard form ratios
- By classification of form on the basis of form ratios
- By compilation of taper tables

(a) By Comparison of standard form ratios

For this purpose two form ratios are:

(1) Form factor and (2) Form quotient

Form factor

Form factor is defined as the ratio of the volume of a tree or its parts to the volume of a cylinder having the same length and cross section as the tree.

In other words, form factor is the ratio between the volume of a tree to the product of basal area and height. Depending on the height of measurement of basal area and on the parts of the tree considered, the following classes of form factors are distinguished.

The form factors may be used for following purposes:

1. To estimate volume of standing trees
2. To study laws of growth

(i) Artificial form factor

The basal area is measured at breast height and the volume refers to the whole tree both above and below the point of measurement. Artificial form factor is also known as breast height form factor.

(ii) Absolute form factor

It is the ratio between the volume of the tree above the point of diameter or basal measurement with the cylinder which has the same basal area and whose height is equal to the height of the tree.

(iii) Normal form factor or true form factor

In this form factor basal area is measured at a constant proportion of the total height of the tree, e.g., $1/10^{th}$, $1/20^{th}$, etc., of the total height and the volume refers to the whole tree above ground level.

This form factor has several disadvantages, viz., (i) the height of the tree has to be determined before the point of measurement can be fixed and (ii) the point of measurement may be very inconvenient in case of very tall as well as very short trees.

Absolute form factor and normal form factor are no longer used. Unless otherwise stated, form factor in India implies artificial form factor with basal area calculated at 1.37 m above ground level. Expressed as an equation, this form factor may be described as:

$$V / Sh$$

Where, F is the form factor

V is the tree volume in cubic units

S is the basal area of breast height in area units

h is the height of the tree in linear units

Kinds of form factor

Depending on the volume represented, form factors are classified into following kinds:

1. Tree form factor
2. Stem timber form factor and
3. Stem small wood form factor

Form height

Form height is defined as the product of form factor and total height of tree. This can be expressed as a formula as follows:

$$\text{Form factor Fh} = \frac{V}{S}$$

Where Fh is form height

V is the volume of the tree and S is the basal area

Form quotient

Form quotient is defined as the ratio between the mid-diameter and the d.b.h.

$$\text{Form Quotient E.Q.} = \frac{\text{Mid} - \text{Diameter}}{\text{dbh}}$$

Classification based on Form ratios

(i) Form class (ii) Form point ratio

(i) Form class

Form class is defined as one of the intervals in which the range of form quotients of trees is divided for classification and use. It also applies to the class of trees which fall into such an interval. Trees may be grouped into form classes expressed by form quotient intervals such as 0.50, 0.55 to 0.60 and so on or by mid-points of these intervals such as 0.525, 0.575 and so on.

(ii) Form point ratio

Form point ratio is defined as the point in the crown as which wind pressure is estimated to be center. Therefore form point ratio (or height) is defined as the relationship, usually expressed as a percentage, of the height of the form point above ground level to the total height of the tree.

C. Classification by compilation of taper tables

Another method of studying form of trees is by compilation of taper tables, which show the actual form by diameters at fixed points from the base to the tip of a tree.

They are used for the following purposes.

(i) Volume of the average tree for each diameter and height class can be found readily without direct measurement. The only measurement that will be needed is the d.b.h. (o.b.) and the height of standing tree.

(ii) Volume tables can thus be prepared from taper tables in desired units.

The following kinds of taper tables are generally met with:

(i) Ordinary taper tables or diameter taper tables

These tables give the taper directly for diameter at breast height without reference to the tree form. Thus they enable computation of volumes without reference to the tree form.

(ii) Form class taper tables

These tables give for different form classes the diameters at fixed points on the stem expressed as percentages of d.b.h. (u.b.).

Measurement of area

The cross-section of the tree is circular, the sectional area can be determined by the formulae:

$$\text{Area} = \frac{\pi d}{4}$$

when d is the diameter of the tree at the section

or

$$\text{Area} = \frac{g^2}{4\pi}$$

when g is the girth of the tree at the section

VOLUME MEASUREMENT

The ultimate aim of any measurement is to determine the quantity of wood in terms of volume either for direct utility or for sales or for management or for using and research purposes.

I. Calculation of volume of felled trees

1. Felling and conversion

The felled trees are converted into logs and the length depends on rate of taper and the market. For example the logs suitable for paper industry is converted into 1m billets and the logs suitable for plywood industry is over 1 m and the log size of standard timber is over 3m. Hence logs are converted into 3 m billets and the top end log may go upto 4.5m. If the top end log is over 1.5 m it may be treated as a separate log.

2. Volume estimation

Due to the biological nature of trees, the felled logs are not real cylinder but look like a cylindrical form and the base always look like a frustum of a neiloid, the middle portion look like a frustum of a paraboloid and the top portion look like a frustum of a cone. Hence various scientists have designed several formulae taking into consideration the form of the felled logs and the formulae are furnished below.

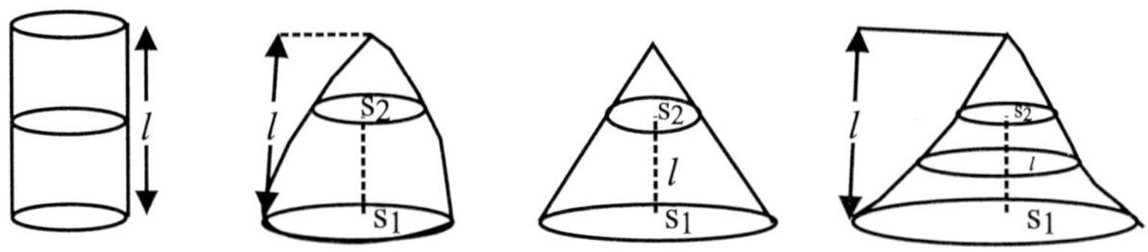

Cylinder **Paraboloid** **Cone** **Neiloid**

Form of solid	Volume of full solid	Volume of frustum of solid	Remarks
Cylinder	Sl	sl	
Paraboloid	sl/2	$(s_1 + s_2)/2 * 1$	Smalian's
Cone	sl/3	$s_m * 1$	Huber's
		$\{(s_1 + s_2 + (s_1 + s_2)^{1/2}\}/3 * 1$	
Neiloid	sl /4	$(s_1 + 4 s_m + s_2)/6 * 1$	Newton's formula

Where *s* is the sectional area at the base in square units

S_1 is the sectional area at the thick end in square units

s_m is the sectional area at the middle in square units

s_2 is the sectional area at the thin end in square units

l is the length of the log or height of the solid in linear units

3. Quarter girth formula

In India, predominantly volumes of logs are calculated using the quarter girth formula:

$$\text{Volume of log} = (g/4)^2 \times l$$

where, g is the girth of the log at the middle and

l is the length of the log

Commercial volume

It is the volume of stem measured down to a thin end diameter upto which conversion is usually done. This volume excludes volume of the stump.

Standard stem timber

It is the volume of stem wood timber in round from ground level down to 20 cm diameter over bark, volume being taken exclusive of bark

Standard stem small wood

It is the volume of stem wood in round between 20 cm diameter over bark and 5 cm diameter over bark, volume being taken inclusive of bark.

4. Volume estimation using timber calculators

The volume of timber is also measured through a specially prepared tables called timber calculators and they are designed based on quarter girth formulae. The logs with a length ranged between 1 and 25 m and the girths upto 300 cm can be assessed through these tables. Similarly sawn timber length of 1-25 m and the cross sectional area of 1 x 1 cm^2 to 25 x 25 cm^2 can also be measured using these tables.

II. ESTIMATION OF VOLUME OF STANDING TREES

The volume of standing trees without felling can be measured through ocular, direct measurement and indirectly through instrumental methods.

(i) Ocular estimate

Traditionally the practicing forester and the commercial timber merchants uses ocular means of estimation and till today this method exists but needs experience and hence this method is not approached for scientific investigation and confirmation.

(ii) Partly ocular and partly by measurement

The problem is ocular estimation can be resolved by deploying instruments for measuring height and diameter and then estimating the volume keeping the tree tapper into conservation.

(iii) Direct measurement

It is similar to volume estimation of felled trees and in this method direct measurements are made at different heights by climbing on to the tree. Due to difficulty and cumbersome this method is not frequently used.

(iv) Indirect measurement

Several instruments like relaskop, pentaprism calliper and other dendrometers are used for estimation of height and diameter and then volume is calculated.

(v) Estimation of volume using volume tables

Volume estimation is also done nowadays through the volume table established for the purpose. Volume tables are now available for many species for use.

CHAPTER - 9

FOREST UTILISATION – WOOD AND WOOD PRODUCTS

Historically, forests have been recognised as a provider of goods and services. The wood has always been a significant product in the human civilization and is believed to be a renewable natural resource. Besides wood, the forests continue to provide wide range of products like gums, resins, fibres, flosses, canes, honey, dyes, essential oil etc. which extend sustainable livelihood opportunities for the original dwellers and the people lived in and around the forest. Hence this chapter elaborates forest utilization under the following sub themes.

- Definition
- Wood Product
- Solid wood
    - Composite wood
    - Non-Wood Forest Products
- Fibres
    - Flosses
    - Grasses
- Bamboos and canes
    - Tan
    - Gums
    - Dye
    - Resin and Oleoresin
    - Essential oil
    - Oil seeds
    - Animal products

DEFINITION

The process of harvesting, converting and disposing off forest produce and other resources of forests is termed as Forest Utilization. Technically, forest utilization includes extraction and felling of wood, conversion, transportation and manufacture of various end products.

The following are the main branches of Forest utilization.

- Felling
- Extraction
- Conversion
- Transport
- Storage
- Marketing
- Value addition

More than 95% of India's forests are under the control of the State wherein logging operations are at the barest minimum. However, the country supports a considerable area under the category "Trees Outside Forests". According to the India State of the Forests Report 2017, the total growing stock of "Trees Outside Forests" is 1603.99 million cu. m. These areas and are categorized as "TOF" have a huge potential to supply timber and industrial wood to cater the needs of various wood based industries of the country.

WOOD PRODUCT

Wood utilization

Wood is used for various purposes from timber for construction, plywood, veneer, pencil, agriculture implements, match, splints, packing cases, sports goods etc. The timber utility varies from species to species and again depending on the industries. The major species which yield wood and wood products are discussed below.

- Wood
- Composite wood

Wood

Wood is anatomically a secondary xylem and chemically it is a lignocellulosic biomass material. The wood is classified into sap wood and heart wood. Sap wood is the outermost portion of the wood which is lighter in colour and consists of living cells which involve in conduction of sap. Heart wood is the inner dark and central portion of wood primarily made up of dead lignified cells and provides strength and rigidity to plant.

Based on wood quality, tree species are classified into hard wood and soft wood. The distinctive characters of hard wood and soft wood are compared in following table.

Hard wood	Soft wood
All hardwood species are angiosperms	All softwood species are gymnosperms
Hardwoods grow in tropical regions of the world	Softwoods grow in temperate regions of the world
They are generally broad leaved	They posses needle leaved in majority of species
Hardwood xylems consist of various proportion of different kind of cells and are thus often uniquely and pectacularly figured. It contains major kinds of cells, each constituting 15 % or more volume of the hardwood xylem.	The xylem of softwood is simple and has only four or five different kind of cells. Among these, one or two occurs in appreciable numbers
Hardwood contains xylem vessels for conduction of water and xylem fibres for the mechanical strength.	90-95% of soft wood xylem is composed of longitudinal tracheids. They take the role of both water conduction and mechanical support.
Hardwood has multiseriate rays. 1-30 cells width and are not aligned in straight line.	Softwood has uniseriate rays. Cells are aligned in straight line.
Most hardwood has a higher density than most softwood.	Most softwood has a lower density than most hardwoods.
Examples of hardwood trees include teak, mahogany, alder, balsa, beech, maple and oak.	Examples of softwood trees are cedar, Douglas fir, juniper, pine, redwood, spruce, and yew.

Utilization of wood

The wood of any tree species has profound utility and it is based on wood physical, chemical and anatomical properties. In general, the woods has several utility and are furnished below.

1. Wood as a structural timber

The solid wood of several species are used as raw material for structural timbers for construction of building, houses and bridges. It can also be used for variety of purposes like beams, doors and windows, floors, panels, roofing and ceiling etc.,

Example: *Acacia catechu, Albizia prosera, Dalbergia latifolia, Artocarpus, Dipterocarpus, Hopea, Largerstromia, Madhuca, Mangifera, Mesua, Pinus, Pterocarpus, Shorea* etc

2. Wood as a raw material for sports goods

Sports goods which are designed as bent wood articles which primarily include hockey sticks, badminton rackets, tool handles, walking sticks, boat wheels, furniture etc. the wood of some of the species are used in manufacturing of the above mentioned bent wood article.

Example: *Dalbergia latifolia*, *Tectona grandis*, Largerstromia, Mangifera, *Toona ciliata*, Stereospermum etc.

3. Wood as a raw material for agriculture tools and implements

Several agricultural implements like ploughs, clod crusher, hand hoe and spade are developed using wood as tool handles. For this purpose only the strongest, toughest and hardest woods are used.

Example: *Acacia catechu, Dalbergia sissoo, Hardwickia binnata, Acacia nilotica, Anogeiscus latifolia, Pterocarpus marsupium.*

4. Wood as a raw material for shoe heels and boot lasts

Some of the species particularly species with medium weight, moderately hard and retain stability after seasoning are used in manufacturing of boot lasts.

Example: *Dalbergia sissoo, Mangifera indica, Pterocarpus marsupium.*

However for shoe heels the wood should be light and moderately tough and should not chip off.

Example: *Boswellia serrata, Bombax ceiba, Holoptelia, Trewia nudiflora.*

5. Wood as a raw material for carts and carriages

Carts and carriages are used in most agriculture and allied activities. Several species with stability character are used in manufacturing various parts of carts and carriages like wheels, spokes axle and main framework.

Example: *Dalbergia sissoo, Tectona grandis, Hardwickia binnata, Acacia catechu, Hopea parviflora, Artocarpus* etc.,

6. Wood as raw material for containers

Several wood containers are used in various business operations and these containers are barrels, tubs, Drums, vats and casks. The wood with straight grain and feasible nature coupled with impenetrable and non shrinking quality are used for this purpose.

Example: *Tectona grandis, Lagerstormia, Terminalia*, Shorea and Grewia.

7. Wood as raw material for furnitures

Wood of several species are used in furniture cabinets and panel making process chairs, tables, sofa sets, billiards tables, screens, TV and radio cabinets. For this purpose the timber with attractive odour, good grain, non liability to crack, split, warp and with easy wood working quality are deployed.

Example. Chukrasia, *Albizia procera, Toona ciliata, Dalbergia latifolia* and *Pterocarpus marsupium*.

8. Wood as raw material for match splints and boxes

Wood forms a critical raw material for manufacturing of match splints and boxes. The wood with straight grain and colour, good fisility, knot free, easy peelability and easy availability is preferred.

Example: *Ailanthus excelsa, Ailanthus triphysa, Anthocephalous cadamba, Populus deltoides, salix, Albizia falcatoria.*

9. Wood as raw material for Musical instruments

Wood plays a vital role in the manufacturing of several musical instruments. For this purpose wood with light weight coupled with good acoustic properties is preferred.

Examples

- Drums : Albizzia, Pterocarpus
- Violin : Acer, Teak
- Veena : Artocarpus, Gmelina
- Sitar : *Toona ciliata*, Teak, *Cedrus deodara*
- Harmonium : Teak
- Guitar : Canarium, Morus

10. Wood as raw material for Packing cases

The wood with light weight, straight grain, free from knots and splits, resistance to termite attack is preferred.

Example:#Mango, Rubber, Albizia, Pinus is preferred

11. Wood as raw material for Railway carriages

Several parts of railway carriages like seats, sleeping birth, interior fittings and etc., are made using solid woods of various species.

Example: Teak, Artocarpus, Dalbergia, Terminalia

12. Wood as raw material for Ship building

Wood is used as raw material for construction of large ships to smaller boats including sail boats and yatches. The timbers with strong elastic and durable characters are preferred for this purpose.

Example: Teak, Dalbergia, Cedrus, Albizia, Melia

13. Wood as raw material for Decorative materials

Wood has several decorative characteristics like color, grain quality, figure, lusture etc., The color and lusture of the both sap and heart wood has excellent decorative and ornamental qualities.

Examples: *Artocarpus, Acrocarpus, Albizia, Adiena cordifolia, Dalbergia latifolia, Mangifera indica, Pterocarpus dalbergioides,* Teak, Shorea and *Xylia xylocarpa*

COMPOSITE WOOD

It is the general term for built up bonded product consisting either wholly of natural wood or of wood in combination with metals, plastics, etc.

The most common composite wood used in India is plywood. The raw material used in plywood industries are

i. Poplars
ii. Eucalyptus
iii. Silver oak
iv. *Melia dubia*
v. *Acrocarpus fraxinifolius*
vi. *Maesopsis eminii*

a. Plywood : It is a glued wood construction built up of veneers in such a manner than the grain of each veneer is at right

Plywood manufacturing process

Step 1: Felling and Transportation

The plantations grown after 5 years are eligible for felling. The minimum girth requirement for felling *Melia dubia* tree is 20 inches girth at breast height. The plantations with the girth of 20 inches and above are felled into bellets of 4.3 ft length uniformly using power chain saw or two men saw. The converted logs are then transported to factory.

Step 2: Peeling of Logs

For peeling logs, the 4 feet peeling lathe is used. The logs with 51 inch length (4 feet & 3 Inch) logs are kept in the Lathe for peeling. While peeling, the core veneers are produced into two different thickness viz., 1.8 mm or 2.0 mm as per the standard requirement. The core veneers are then cut into 2 feet length through the chopping machine. The logs are normally in green state with moisture content ranged between 30 and 44% and are directly deployed for peeling and hence all the peeled veneers are with very high moisture content.

Step 3: Drying of Veneers

The veneers produced have higher moisture content and for making plywood the optimal moisture requirement is 8-12%. Hence the veneers have to be dried for which a drier is used. The core veneers are fed into the dryer wherein the temperature is maintained at 120°C. The drying process ensured the moisture content of around 12%.

Step 4: Gluing

After drying, the veneers are sized into uniform sizes and are intended for gluing. For this purpose, two types of resins are used. Melanine Formaldehyde resin (MF resin) is used for gluing for the process of commercial plywood production and phenol formaldehyde resin (PF resin) is used for gluing towards the production of water proof plywoods. For the purpose of gluing, a glue spreader machine is used. Using the glue spreader machine, all the core veneers are glued either by MF Resin or PF Resin.

Step 5: Construction of Plywoods

After gluing, the plywoods are constructed using both face and core veneer. The face veneer is used from *Dipterocarpus turbinatus*. In the plywood construction process, first one face veneer is laid and upon the face veneer, glued core veneers are kept horizontally and vertically to the required thickness and again another face veneer is laid on the assemble of core veneers. After the assemblage of face veneer at two ends with core veneers in the middle and the whole process is called plywood construction.

Step 6: Pressing Plywoods in the Hot Press

After construction, the plywood construct has been placed in a hot press for pressing. It is done through a hot press having a hydraulic compressing capacity of about 1000 tonnes. In the hot press, the temperature of about 140°C is maintained to ensure uniform pressing. The assembled plywood is kept inside the hot press for about 20 minutes. During this process, the glue applied on the core veneers is spread uniformly and bonding of the core veneers takes place very solidly due to the hydraulic compression and the hot press plywood is ready for further process.

Step 7: Cutting Process

The hot dried plywood constructs have uneven sizes which need to be sized in all directions. For this purpose, the DD saw cutting machine is employed and the four edges of plywoods are cut in order to make the required sizes of plywood. This cutting process is preferred to ensure uniformity and to remove the uneven veneer edges.

Step 8: Sanding

After the cutting process, the plywood needs to be sanded to convert rough surface into smooth surface. For this purpose, a sanding machine is used in which both sides of plywoods are sanded. During this process, the plywood becomes very smooth on both faces of the plywood and are ready for supply to consumers.

Step 9: Labelling

After the sanding process, the plywoods are labelled and are ready for despatch.

Step 10: Loading and Transportation

The required grades of plywood are loaded and transported to the dealers and retailers through trucks on a regular basis

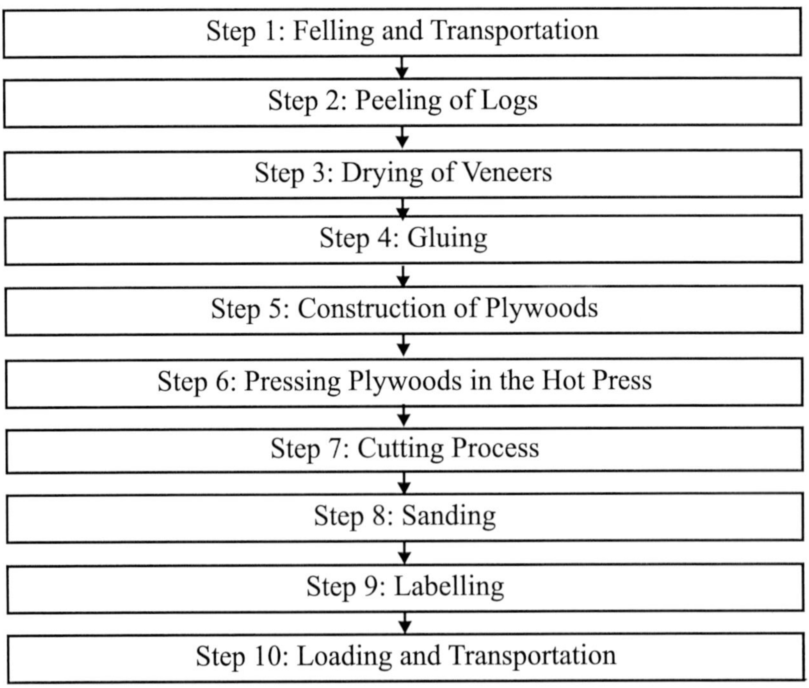

b. Laminated wood

These are a build-up product made of wood layers (laminate) in which all layers laid with their grain parallel and glued or otherwise fastened together. The glued laminated wood is called as glulam. Lamina refers the thin board of veneers. Lamina are seasoned in kilns and cut into uniform width and length. These are arranged in a proper order. This feed into a glue spreader and assembled into a required shape. Then pressure is applied in a clamp fitting assembly.

c. Core boards: These are composite board built up of a core composed of strip of woods of various dimensions glued together to form a slab. This is intern glued between two or more outer veneers with the direction of grain of the core strips running perpendicular to that of adjacent veneer.

Board	Width
Bottom board	7.5cm
Black board	<2.5 cm
Lamin board	<7 cm

d. Sandwich board: It is the general term for built up boards having a core of light material, faced on both sides with relatively thin layers of woods having high strength properties. The thin faces are called as skins which are strong enough, have high strength and more stiffness. It is used in manufacturing parts of air craft, motor boats, table tops and containers.

e. Fibre board: It is a sheet of material made from fiber of wood. Fiberboards are made from all wood wastes and agricultural wastes. First the wood is defibriated or pulped inter-felted into a mat. Bonding agents and supplementary agents added at the felting stagc to improvc thc mcchanical properties. This is consolidated by pressured and heat. It is used as a core material in sandwich and core boards, building boards, insulation boards and wallboards. Based on densities, these are classified as (i) Hardboard (ii) Semi hardboard (iii) Medium hard board (iv) Soft board.

f. Particle board: It is a board or sheet constituted from the fragments of wood and other lignocellulosic materials bonded with organic binders with the help of one or more agents like heat, pressure, humidity, catalysts, etc. This is made from chips, flakes, splinters. The main types are (1) Chip board (2) Flake board and (3) Shaving board

Waste materials are chipped in chippers and screened where unwanted materials are removed. This is dried to suitable moisture content in a drier. Mechanical mixers were added and flat pressed by hot process method.

g. Improved wood : These are also called as modified wood. It is a general term of wood that has been specially treated in various ways to reduce or retard working or to increase strength.

The wood properties such as hygroscopicity, strength, density etc., are modified by impregnating with resins or by compression or by heating.

h. Impregnated wood: It is a wood impregnated with impregnants. The common impregnants used in this process are wax, paraffin, resins and oils. The main objectives are to make the wood dimensionally stable and prevention of water absorption. Now commercial water soluble, thermosetting formaldedyde resins is available. In the manufacturing process, first woods are impregnated through vacuum impregnation with or without pressure. This is thermosetted and impregnated again. After curing, individual materials are bonded together as one.

i. Heat stabilized wood: This is obtained by simple heating. In this, woods are passed beneath the surface of a molten metal or fused salt. It is subjected to 260 – 315°C for few minutes under non oxidizing conditions.

j. Compressed wood: In this, woods are compressed without impregnation. It increases specific gravity and strength of the wood so that it can withstand high wear and tear. Compressed wood is formed from solid or laminated wood under the pressure of 90 – 140 kg/cm^2. Material obtained after compression have a specific gravity of 1.2 to 1.3.

k. Compregnated wood: This designates the wood which are subjected to impregnation and compressed. Woods are spliced into veneers and impregnated with urea formaldehyde by compression. This increases tensile strength and improves insulation property. Because of this treatment, wood becomes 20 times harder than normal woods.

l. Heat stabilized wood: In this woods are subjected to heat to improve wood property.

m. Chemically modified wood: In this wood, OH group of cellulose is replaced. Chemicals used for this are acetic anhydride influences of pyridine as catalysts. Urea hydroxylin and sulphuric acid also used. It decreases the hygroscopicity and also of swelling and shrinkage.

Utility of composite woods

- Composite woods are useful in construction industries
- They are flexible, uniform and consistency in performance and they are used in all interior decorations.
- In multistory buildings the composite wood are preferred due to its light weight and durability.
- Some of the composite woods are stronger and more dimensionally stable and hence they are used in construction of wood houses.
- Composite woods are also used to make headers, beams, girders, columns and also for heavy trusses.

FOREST UTILIZATION - NON WOOD FOREST PRODUCTS

Non Wood Forest Products includes all kinds of forest produce except timber and firewood. They comprise all vegetable, animal and mineral products found in forest areas. The important non wood forest products of Indian forests are furnished hereunder.

A. Fibres,

B. Flosses,

C. Grasses

D. Bamboos and canes,

E. Tan,

F. Dye,

G. Gum

H. Resin and Oleoresin

I. Essential oil

J. Oil seeds

K. Animal products

A. Fibres

The phloem, or bast, of both gymnosperms and dictoyledons consists of among others, long, thick walled (fibrous) sclerenchymatous tissues. The bast fibres are not lignified like the fibres of wood and are therefore, of commercial value for the manufacture of ropes and cordage. The leaves, left stalks and fruits of monocotyledons yield to tough and flexible stands of fibre.

Uses of fibres

- Fibres are used for cordage and in the textile industry (the manufacture of fabrics and netting). Twine, fish lines, ropes and cables are among the many kinds of cordage. Fabrics include cloth for wearing apparel, soils etc and also coarser materials, such as gunny and coarse canvas.
- Tough and stiff fibers are utilized in the manufacture of brushes and brooms.
- Pliable fibres are used in making straw hats, baskets etc.
- Elastic fibres are roughly woven together for matting.
- Fibres are used also for stuffing mattresses, cushion etc.,
- Fibres are used for caulking the seams in boats,
- Fibres as stiffening in plaster and as packing material and
- Fibres are used in the manufacture of paper.

Fibre separation process

Fibres are separated from each other by many methods and the process of separating is called retting. Some bark requires soaking in water for a time and then beaten or scrapped to clean.

Classification of fibres

The fibres may be classified according to the part from which they are obtained as stem and leaf fibres. The majority of plants yielding useful fibres belong to the families Streculiaceae, Tiliaceae, Leguminosae, Asclepiadaeae and Urticacease.

S.No	Common Name	Scientific Name
I. Stem Fibre		
1.	Elephant rope tree	*Sterculia vilosa*
2.	Indian screw tree	*Helicteres isora*
3.	Dhaman tree	*Grewia tiliifolia*
4.	Flame of the Forest, Bengal kino, Palas tree	*Butea monosperma*
5.	Bando Lata	*Butea parviflora*
6.	Crown flower	*Calotropis gigantea*
7.	Bark cloth tree, Upas tree	*Antiaris toxicaria*
8.	Indian charcoal-tree pigeon wood	*Trema orientalis*
9.	Marijuana, Hemp, Gallow grass	*Cannabis sativa*
10.	Himalayan nettle, Nilgiri nettle	*Girardinia diversifolia*
11.	Sea / Beach hibiscus, Coastal cottonwood	*Hibiscus tiliaceus*
II. Leaves Fibre		
12.	Screw pine, Palau Fruit	*Pandanus odoratissimus*
13.	Pandanus Palm, Australian Screw Pine	*Pandanus tectorius*
14.	Umbrella tree, Aromatic Screw tree	*Pandanus fascicularis*
15.	Fishtail palm	*Caryota urens*
16.	Palmyra palm	*Borassus flabellifer*
17.	Wight's Sago Palm	*Arenga wightii*
18.	Sentry plant, Century plant	*Agave americana*
19.	Caribbean agave	*Agave angustifolia*
20.	Aloe	*Aloe barbadensis*
21.	Pineapple	*Ananas sativus*

B. FLOSSES

There are several forest trees and plants which produce flosses (rough silk) in their fruits.

Kapok (Tree cotton)

The kapok of commerce is the fruit hair of certain species of the genera *Bombax* and *Eriodendron* in the family Bombacaceae. The Indian Kapok is derived from the species of Bombax genus. The true kapok (Kapok is Java) is the produce of *Eriodendron anfractuosum* alternatively known as *Ceiba pentandra*. Kapok is known for light elastic, buoyant and has a smooth and shining surface. It does not get wet on account of its water repellent properties.

Uses of flosses

- Indian Kapok is chiefly used for stuffing pillows, mattresses and cushions.
- It is used in the manufacture of life-belts and other naval life-saving appliances.
- For upholstery, kapok is generally preferred to all other filling fibres.
- Due to its non conducting property, it is used for insulation and for interlining in warm clothing.
- Due to its lightness, kapok is used in the preparation of aviator's coat and also for mattresses for the army.

S.No	Common Name	Scientific Name
1.	Yellow Silk Cotton Tree	*Cochlospermum gossypium*
2.	Buttercup Tree	*Cochlospermum religiosum*
3.	Stripped Snake plant	*Sansevieria zeylanica*
4.	Crown flower	*Calotropis gigantea*
5.	Rubber Bush, French cotton	*Calotropis procera*

C. GRASSES

1. Grasses: From the utilization point, grasses may be broadly divided into three groups *viz.*,

(i) Fodder grasses

(ii) Fibre yielding and thatching grasses and

(iii) Oil yielding grasses.

(i) Fodder grasses

Eriochloa polystachya, Sporobolus indicus, Sporobolus diander, Ischaemum pilosum, Andropogon pertusus, Cenchrus ciliaria.

(ii) Fibre yielding and thatching grasses

None of the Indian grasses supply fine textile fibres but a few are suitable for cordage, matting and paper making.

S.No	Common Name	Scientific Name
1.	Sabai grass	*Eliolopsis binata*
2.	Munj grass	*Saccharum munga*
3.	Hardy sugar cane grass	*Saccharum arundinaceum*
4.	Bahia lovegrass	*Eragrostis cynosuroides*
5.	Elephant grass	*Typha elephantine*
6.	Thatch grass	*Saccharum spontaneum*
7.	Satintail grass	*Imperata arundinaceae*
8.	Perennial tussock grass	*Heteropogon contortus*

(i) Oil Yielding Grasses

S.No	Common Name	Scientific Name
1.	Vetiver	*Vetiveria zizanioides*
2.	Palmarosa	*Cymbopogon martinii var motia*
3.	Ginger grass	*Cymbopogon martinii var sofia*
4.	Citronella grass	*Cymbopogon nardus*
5.	Lemon grass	*Cymbopogon flexuosus*

D. BAMBOOS AND CANES

Some of the important bamboos are as under:

***Arundinaria falcate*:** This grows gregariously forming an an undergrowth in forests of oak and deodar, usually in moist shady places at situations between 1,200-2,000 m elevation. The culms are 1.84.0 in high and are used for basket making, fishing rods and hookah pipes.

***Arundinaria spathiflora*:** A small bamboo found growing in Western Himalayas at 2,000-2,700 m in elevation as an under growth of fir, spruce and deodar forests.

***Bambusa arundinacea*:** A thorny bamboo, having culms 24-30 m in height and 15-18 cm in diameter. It occurs wild, especially in the hill forests of western and southern India, ascending upto 900 m. The bamboo is used in paper manufacture.

***Dendrocalamus giganteus*:** This is the biggest of the Indian bamboos cultivated in parts of West Bengal, Assam and Malabar. It is 24-30 m high and 20-25 cm in diameter.

***Dendrocalamus strictus*:** This occurs throughout India, in deciduous forests often of a dry type, on hill-slopes upto 1,000 m elevation. The calms are usually quite solid, 6 × 15 m high and 2.5-7.5 cm in, diameter. It is most universally used of all the bamboos. It is put to a great variety of uses.

***Dendrocalamus hamiltonii*:** This occurs in the sub-Himalayan tract upto 900 m. The culms are upto 25 m long, 10-18 cm in diameter, and thin-walled.

***Dendrocalarnus longispathus*:** This occurs in parts of West Bengal and Assam, usually along streams upto 1,200 m. The culms are from 1248 m in height and 7,5;10 cm in diameter.

***Melocanna bambusoides*:** This bamboo is found in Garol Khasi and Lushai Hills in Assam. The culms are upto 2 m in length and 9.5 cm in diameter. It is typically gregarious.

***Ochlandra travancorica*:** This is found in the mountains of Kerala, Karnataka and Tamil Nadu. It is an erect reed-like, gregarious bamboo, which covers large areas.

***Oxytenanthera nigrociliata*:** This is a tufted bamboo with culms 9-15 m high and 1.3-10 cm in diameter, found in Assam, Orissa and Andaman Islands.

Other Bamboos: *Bambusa balcooa, Bambusa tulda, Cephalostachyum pergracile, Gigantochloa macrostachya, Ochlandra scriptoria, Oxytenanthera ritcheyi, Pseudostachyum polymorphum* are of importance.

Uses of Bamboos

I. Pulp and paper

Bamboo is the main suitable fibrous raw material, found growing extensively in the forests of the country. There is a very large number of species but the main which are used for paper making are: *Bambusa arundinacea, B. nutans, B. balcooa, B. tulda, B. vulgaris, Dendrocalamus strictus, D. hamiltonii, Melocanna barnbusoides and Oxytenanthera monostigma, Ochlandra rheedii* and *O. travancoria*. Bamboo is used for almost all the common varieties of paper such as writing, printing, wrapping paper and also as part of furnish in newsprint.

II. Furniture industry

Bamboos are the major raw material for furniture industries across the world and chineese are leading the business. The bamboo based furnitures are preferred by the consumers than the wood based due to its appearance, easy handling and plasticity. Wide range of furnitures like chairs, tables, stools, doors, huts etc., are made using wide range of bamboo species. The species like *Dendrocalamus strictus, Bambusa bambos, Bambusa balcooa, Dendrocalamus brandsii* etc. are playing a vital role in furniture industry.

III. Bamboo based biomass generation

There is growing interest towards biomass based power generation particularly using bamboo as a raw material. For every megawatt of biomass power generation, it is estimated that 1.5 tonne of bamboos are required. Bamboo is amenable for power generation both through combustion and gasification technology. One of the bamboo species like *Bambusa balcooa* has been used for this purpose due to high calorific value, combustion efficiency and low ash content. Hence the corporate involved in biomass power generation are deploying bamboo as a raw material due to its fast growth and coppicing ability.

IV. Bamboo based activated carbon

The industries involved in activated carbon production started deploying bamboo as a raw material. The activated carbon generated from bamboo has found multifarious utility from filter paper, paints and varnishes, pharmaceutical industries, colouring industries etc., and play a significant role in corporate sector.

V. Important Rattan species

S.No	Genus	Distribution
1.	*Calamus*	Tropical Africa, India and Sri Lanka, China, south and east to Fiji, Vanuatu and eastern Australia
2.	*Calospatha*	Endemic to Peninsular Malaysia
3.	*Ceratolobus*	Malay Peninsula, Sumatra, Borneo, Java
4.	*Daemonorops*	India and China to westernmost New Guinea
5.	*Korthalsia*	Indo-China and Burma to New Guinea
6.	*Laccosperma*	Humid tropical Africa
7.	*Myrialepis*	Indo-China, Thailand, Burma, Peninsular Malaysia and Sumatra
8.	*Plectocomia*	Himalayas and south China to western Malaysia
9.	*Plectocomiopsis*	Laos, Thailand, Peninsular Malaysia, Borneo, Sumatra

D. TANS

Tannins are organic compounds, chiefly glucosoidal in nature which have an acid reaction and are very astringent. They are found in solution in the cell sap and occur in all parts of the plants (chiefly in the parenchymatous – tissues such as bark and young wood certain fruits and leaves and sometimes in the galls formed on leaves) but the amount is very variable. There are only few plants in which they are plentiful.

Tannins are of economical interest because of their ability to combine with albumen and gelatin such as those in animals skins to form an insoluble compound which will resist decay. For this reason raw animal hides treated with tannin become converted into a strong, flexible and resistant material known as leather. Tannic acid can turn iron salts black, grey, purple or green and this property of tannin is made use of in the manufacture of ink.

Tannins may be classified according to the part from which they are obtained as tans, fruit tans and leaf tans-

(a) Bark tans

S.No	Common Name	Scientific Name
1.	Black wattle	*Acacia mearnsii*
2.	Green wattle	*Acacia decurrens*
3.	Silver wattle	*Acacia dealbata*
4.	Golden wattle	*Acacia pycnantha*
5.	Avaram senna	*Cassia auriculata*
6.	Golden rain tree	*Cassia fistula*
7.	Babul	*Acacia arabica*
8.	Sal	*Shorea robusta*
9.	Arjuna tree	*Terminalia arjuna*

Other tan barks

Barks of certain Mangroves like *Rhizophora mucronata* give tanning agents of commercial value. Besides the above there are quite a number of trees whose bark contains appreciable amounts of tannin. A few among them are: *Emblica officinalis, Acacia suma, Zizyphus glaberrima, Lannea coramandelica, Syzygium cumini, Lagerstroemia parviflora, Terminalia tomentosa, Anacardium occidentale, Xylia xylocarpa* and *Careya arborea*

(b) Fruit tans

Fruits of certain trees are used for tanning purposes of which the most important are the

S.No	Common Name	Scientific Name
1.	Chebulic myrabolan	*Terminalia chebula*
2.	Beleric myrabolan	*Terminalia bellerica*

S.No	Common Name	Scientific Name
3.	Emblic myrabolan	*Emblica offcinalis*
4.	*Babul*	*Acacia arabica*
5.	Kath Ber	*Zizyphus plaberrima*
6.	Midnight horror, Indian trumpet Flower	*Oroxylum indicum*
7.	Indian persimmon	*Diospyros peregrina*
8.	Divi-divi	*Caesalpinia coriaria*

(c) Leaf tans

Leaves are not important tan agents. However leaves of certain plants can be used for tanning.

S.No	Common Name	Scientific Name
1.	Axle wood tree	*Anogeissus latifolia*
2.	Karunda	*Carissa spinarium*

Uses of tannins

- Tannins are used in leather industry as tanning material.
- It is used in several leather industries like belt, shoes, chapels and suitcase etc.
- Tannins are used in upholstery industry.
- Sports good industry also uses tannins.
- It is also used in pharmaceutical industries
- Several Aurvedic medicines uses tannins
- Some of the tannins are used as mordant in dying process.

F. DYES

Dyes are yielded from certain barks, wood, roots, leaves, flowers, fruits and seeds. Many tanning barks and wood-extracts are also used for dyeing. Some plants yielding colourings are used in dyeing in the form of mordents to fix the dyes from other plants.

Uses of dyes

- Dyes are used in colouring process particularly in paints, varnishes, leather, and paper industries.
- The natural dyeing process in textile industry prefers vegetable dyes.

- Some of the dyes are useful in food and beverage industry as replacement to chemical dyes.
- The dye obtained from red sanders is used for dyeing leather and staining wood. Vegetable dyes used in calico printing.
- In pharmaceutical industry, dyes are used in colouring the capsules.
- Dyes obtained from several species are used in wide range of cosmetic industries.
- The colouring process in ice creams and confectionaries also use vegetable dyes.

Dyes and suitable species

Bark dyes	**Wood dyes**	**Root dyes**
Mimusops elengi	*Pterocarpus santalinus*	*Morinda tinctoria*
Lannea coromandelica	*Caesalpinia sappan*	*Punica granatum*
Casuarina equisetifolia	*Acacia catechu*	*Oldenlandia umbellate*
Terminalia tomentosa	*Artocarpus heterophyllus*	*Morinda citrifolia*
Cassia alata	*Artocarpus lakoocha*	
	Adenanthera pavonina	

Flower dyes	**Leaf dyes**	**Fruit dyes**
Butea monosperma	*Terminalia catappa*	*Mallotus Philippinensis*
Cedrela toona	*Wrightia tinctroria*	*Bixa orellana*
Michelia champaca		*Thespesis populnea*
		Wringhtia tinctoira

G. GUMS

Gum is a more or less viscous substance which exudes from cracks or wounds in the bark of many trees shrubs and climbers. Gums are polysaccharides of natural origin, capable of causing a large increase in a solutions viscosity, even at small concentration.

Process of gum secretion

True gums are formed as the result of the disintegration of internal tissues, for the most part from the decomposition of cellulose, through a process known as gummosis. Gums chiefly occur in the cortex tissues. They contain a large amount of sugar and are closely allied to the pectins. They are soluble in water, either dissolving completely or swelling upto form a mucilage, but are insoluble in alcohol and other (unlike resins). They exude naturally from the stems especially when bark cracks with prolonged heat or in response to wouding. Commercial gums reach the market in the form of dried exudations. Gums are specially common in plants of dry regions.

Uses of gums

- Gums are predominantly used in adhesive industry
- They find profound utility in calico printing
- The gums are used in paints and other bonding materials
- Gums are also used in food industry
- Gums are used in pharmaceutical industry.

(i) True gums

S.No	Trade Name	Scientific Name
1.	Indian Gum Arabic	*Acacia Arabica*
2.	True gum Arabic	*Acacia senegal*
3.	Brittle gum or Jhingan gum	*Macaranga beltata*
4.	Lannea gum	*Lannea coromandelica* (*L. grandis*)
5.	Amber gum	*Azadirachta indica* (Neem)
6.	Khair gum	*Acacia catechu*
7.	Dhaura gum	*Anogeissus latifolia*
8.	Reddish brown gum	*Aegle marmelos*
9.	Anacardium gum	*Anacardium occidentale*
10.	Feronia gum	*Feronia limonia*

(ii) Pseudo-gums

Gum which is not soluble in water but absorbs water readily and swell up into a jelly mass.

From '*Astragalus heratensis*' a Perisan shrub, exudes, a gum called 'Katira'. This is known in the Indian commerce as 'Gum Tragacanth'. The following are the Indian substitutes (Indian katira or bassora gums) for this gum.

S.No	Trade Name	Scientific Name
1.	Karava or Kullu gum	*Streculia urens*
2.	Dark brown gum	*Bombax ceiba*
3.	White gum	*Moringa oleifera*

S.No	Trade Name	Scientific Name
4.	White gum	*Cochlospermum religiosum*
5.	Pectin gum	*Saccopetalum tomentosum*

(iii) Astringent gums

S.No	Trade Name	Scientific Name
1.	Gum Kino	*Pterocarpus marsupium*
2.	Bengal Kino	*Butea monsperma*

H. RESINS

Resins represent oxidation products of various essential oils and are very complex and varied in their chemical composition.

Uses of resins

- Resins are used in varnishes. Their ability to harden gradually makes commercial varnishes possible. Another property of resins of industrial importance is their ability to dissolve in alkalies to form soap.
- Resins are also used in medicine
- For sizing paper
- As a stiffening material for mats
- In the preparation of sealing wax, incense, perfumes etc.

Resins are broadly classified into two classes

- Pine resins
- Resins from broad-leaved species

Pine resins

Pine resins are obtained from *Pinus roxburghii*, *Pinus wallichiana* (*P. excelsa*) and *Pinus insularis* (*P. khasya*).

Of these, *Pinus roxburghii* or chir pine is by far the most important from a commercial point of view and systematic tapping of *P. roxburghii* alone has been carried out in India.

The resin occurs in resin ducts which run in vertical and horizontal direction through the wood. The resin in the sapwood flows more copiously than that in heartwood, where the resin ducts are often blocked up.

Resin-tapping

When a tree is blazed a number of resin ducts will be broken and the liquid resin contained therein will ooze out. This is the theory of practical resin tapping employed universally. But the actual details may vary.

Method

- The outer bark is first removed where the blaze is to be made (South side) for a space of about 2 feet wide from the bottom to about 3 to 3 feet ups the tree, with an axe. This facilitates the cutting of the blaze.
- A curved cut, about 8 inches broad, is made at a height of about 6 inches from the ground. Into this is admitted as trip of galvanized iron 6" long by 1½ to 2 inches wide which slopes slightly downwards and acts as a slip to guide the flow of resin into an earthenware cup, 6 inches high and 4 inches broad at the top, fixed beneath.
- The channel is then made with a special adze perpendicularly above the lip, leaving about 1½ inches between the chisel-cut for the lip and the bottom of the channel.
- The channel is about 6 inches to one foot long, 4 inches wide and about ½" deep, the greatest depth being near the bottom of the blaze.
- The resin flows down the blaze and is collected in the cup.
- The blaze is freshened once in 6 days i.e. a minimum of 4 freshening are made in a month. The freshening consists in paring off with the adze a thin shaving of wood, so as to open up the clogged resin ducts.
- In this work the blaze must not be lengthened by more than ½ inch at each freshening.
- At the end of the first season the blaze becomes 2 feet or more in length.

Methods of tapping

Two different methods of tapping are adopted.

(i) Light continuous tapping

In this method all trees between 3 feet 6 inches and 7 feet are given one channel and those over 7 feet two channels. The first channel is cut on the southern face as this is the warmest and gives the best outturn of resin. After 5 years' tapping a new channel is started 4½ inches to the left of the old channel. Again after 5 years tapping the third channel is cut 4½ inches to the left of the second and so on till tapping has proceeded right round the tree. When the tree has been gone round in this manner channels are started on the inner-channel spaces of 4½ inches left untapped. The same method is adopted when two channels are placed 4½ inches apart (for trees above 7 feet in girth). The tree is kept always under tapping in this method. Advantages simple and gives a sustained yield.

(ii) Heavy tapping

It is slaughter tapping and is adopted in areas where the trees are to be felled within 5 years. As many channels as possible are out in the tree to obtain the greatest possible cut turn of resin for the 5 years previous to being felled.

Yield

With moderate tapping an average tree may yield from 5 to 12 lbs. of crude resin per annum. The crude resin is stringed through sieves to remove pine needles and other impurities and then transported to distillery in closed tins.

The manufacture of resin

The crude pine resin consists of two principal constituents, (i) Oil of turpentine, a liquid and (ii) resin or colophony, a solid at ordinary temperature. If crude resin be heated with a little water to a temperature of approximately 158°C the oil of turpentine distills out and the resin remains behind. The turpentine is redistilled. One maund of crude resin of *Pinus roxburghii* produces an average 1½ gallons of turpentine and ¾ maund of colophony.

Uses

- Turpentine: Used in the preparation of paints and varnishes, in medicines especially for sprains and rheumatism. Also used for dissolving fats resins and rubber.
- Resin: Used chiefly in soap making, for sizing paper, soldering, in the manufacture of seasling-wax, varnishes, ointments, oil clothes, inks etc. Also used as a drier in the manufacture of paints, varnishes and as dressing for machine belts and bows of violins.

Resins from broad-leaved species

These resins are usually known under the trade name of 'Dammar'. True dammar is obtained from '*Agathis loranthifolia*', a conifer found in Borneo and Sumatra.

Dammars are chiefly used in spirit varnishes and in the manufacture of nitrocellulose lacquers. Dammar varnishes are particularly well adopted for varnishing paper because their luster and light colour.

S.No	Trade Name	Scientific Name
1.	White dammar	*Vateria indica*
2.	Black dammar	*Canarium strictum*
3.	Sal dammar	*Shorea robusta*
4.	Rock dammer	*Hopea odorata*
5.	Green dammar	*Shorea tumbuggia*

Gums resins

As the name indicates, gum resins are mixtures of both true gums and resins and naturally combine the characteristics of both groups. They often contain small amounts of essential oils and traces of colouring matter. Gum resins occur naturally as milky exudations and collect in the form of tears or irregular masses. They are produced by the most part of plants of dry, arid regions, especially species of the natural order. Umbelliferae and Burseraceae.

These may be classified as

1. Emulsive (Gamboge)
2. Foetid (Asafoetida)
3. Fragrant

(1) Emulsive (Gamboge)

Gamboge dye of commerce is a gum resin exuded from the Siamese gamboges tree viz., *Garcinia hanburyi. Garcinia Morella* approaches the Siamese Gamboge tree. Its resin is used to colour silk yellow and for water colour paints.

(2) Foetid (Asafoetida)

Not of forest importance.

Ferula asafetida

Roots send up flowering stem which is cut off and the juice is collected. It is a thick opaque gum resin and is the asafetida of European Commerce.

(3) Fragrant

Fragrant gum resins are used as incense during religious observations and in manufacturing perfumes and fumigating powders.

Boswellia serrata

The Indian olibanum yields a gum-resin called 'Salai gum' which is burnt as an incense. Also used in making up ointments, especially for rheumatism.

Ailanthus malabarica : (Ailanthus triphysa)

Yields a resin used as incense. It is very fragrant.

Oleo resins

The oleo-resins contain a considerable amount of essential oils in addition to the resinous materials. Consequently they are more or less liquid in nature. The essential oils are secreted in the intercellular spaces where they combine with resin formations to produce the oleo-resins which exude from the stem. They have a distinct aroma. They are also known as **wood oils.**

S.No	Trade Name	Scientific Name
1.	*Dipterocarpus turbinatus*	Gurjun oil
2.	*Dipterocarpus tuberculatus*	In-oil
3.	*Dipterocarpus indicus*	Malabar Gurjun oil
4.	*Hardwickia pinnata*	Balsam oil

I. ESSENTIAL OIL

They are volatile oils occurring naturally in various plant parts which include wood, leaf, root, grass, flower and seeds. The species used in production of various sources of essential oil and their classification are furnished below

S.No	Trade name	Species yielding oil
I. Wood oil		
1.	Agar oil	*Aquillaria agallocha*
2.	sandal wood oil	*Santalum album*
3.	Turpentine oil	*Pinus roxburghii*
4.	Deodar oil	*Cedrus deodara*
II. Leaf oil		
5.	Citriodora oil	*Eucalyptus citriodora*
6.	Cinnamom oil	*Cinnamomum zeylanicum*
7.	Eucalyptus oil	*Eucalyptus globules*
8.	Mint oil	*Mentha species*
9.	Geranium oil	*Pelargonium graveolens*
III. Grass oil		
10.	Citronella oil	*Cymbopogon nardus*
11.	Ginger grass oil	*Cymbopogon martini*
12.	lemon grass oil	*Cymbopogon flexuosus*
13.	Palmarosa oil	*Cymbopogon martini var. motia*
14.	Vetiver oil	*Vetiveria zizanioides*
IV. Root oil		
15.	Costus oil	*Saussurea lappa*
16.	Indian Valerian oil	*Valeriana jatamansi*

S.No	Trade name	Species yielding oil
17.	Curcuma oil	*Curcuma aromatica*
V. Flower oil		
18.	Jasmine oil	*Jasminum officinale*
19.	Rose oil	*Rosa damascene*
VI. Seed oil		
20.	Muskdana oil	*Abelmoschus moschatus*

J. OIL SEEDS

Oils are fatty acids which are present in seeds of wide range of species. The seeds which contain appreciable quantity of oils are designated as the tree borne oil seeds (TBO's) and are used for expelling oil through crushing process. Indian forest supports over 100 different tree species which are accredited with oil yielding properties. The important oil yielding species are

1. *Azadiracta indica* (Neem)
2. *Pongamia pinnata* (Pungam)
3. *Calophyllum inophyllum* (Punnai)
4. *Madhuca latifolia* (Madhuca)
5. *Jatropha curcas* (Physic Nut)
6. *Salvodara olllioides*
7. *Shorea robusta* (Sal)
8. *Mesua ferrea*
9. *Simarouba glauca* (Paradise tree)

Uses of TBO's

- The oil obtained from tree species are directly used for edible purposes.
- Seeds of some of the species yield non edible oil but used in soaps, paints and varnish industries.
- The oil obtained from some of the species is used as substitute for butter.
- The non edible oil seeds are used as raw material for biodiesel production after transesterification process.
- Oils of several species are used in siddha and Aurvedic system of medicine
- Oils of several species are used in lighting lamps in temples.
- Oils from some f the species are used in leather tanning industries.

K. ANIMAL PRODUCTS

Animals play an important role in human life. The animals and birds attract people by their activities and their role in providing useful products to human kind. The important animal product which attract human beings are listed in the Table 1.

S.No	Animal products	Fauna yielding the products
1.	Horns	• Sambar deer • Spoted deer • Rhinoceros
2.	Ivory	• Elephants
3.	Hides and skins	• Buffaloes, Sheep, Goat, Deer, Crocodile, Snake and Lizards
4.	Edible bird nests	• *Collaclia inexpectata*- swifts
5.	Silk through sericulture	• *Morus* spp. (Mulberry silk) • *Terminalia arjuna* (Tassar silk) • Ricinus communis (Eri silk) • *Terminalia arjuna* (Muga silk)
6.	Lac through Lac culture	• *Laccifera lacca* • *Butea monosperma* (Ranjeni), • *Schleichera oleosa* (Kusum)
7.	Honey through apiculture	• *Apis dorsata* (Rock bee) • *Apis cerana indica* (Indian hive bee) • *Apis florea* (Little bee) • *Apis mellifera* (European or Italian bee) • *Melipona irridipennis* (Dammer bee or stingless bee)

■■■■

CHAPTER - 10

AGROFORESTRY SYSTEM

Agroforestry is one of the traditional land use systems practised since time immortal. However the organised practice of agroforestry is of recent origin and has gained significant attraction across the world. The growing demand for wood and wood products coupled with restricted supply of wood from natural forest has attracted more and greater productivity form agroforestry. The growing concerns on global climate change and the associated environmental implications also attracted the implementation of agroforestry. All these necessitated incorporation of agroforestry as a subject in several under graduate programme in SAU's and hence this chapter has been outlined with the following sub titles.

- Definition
- Classification and types
  - Agrisilviculture
  - Silvipasture
  - Agrisilvipasture
  - Other systems
- Benefits and Limitation
- Tree crop interaction
  - Above ground interaction
  - Below ground interaction
- Nutrient cycling
- Agroforestry systems for different ecological zones
- Industrial Agroforestry
- National Agroforestry Policy, 2014

DEFINITION

In general, the term Agroforestry means growing trees in combination with agricultural crops. This is an age-old practice which has witnessed its application across the world. Agroforestry has been defined by many researchers and organizations differently but holistically it represents a diverse and sustainable land use system. Some of the definitions proposed are furnished below.

a. Agroforestry is a collective name for all land use systems and practices where woody perennials are deliberately grown on the same land management unit which has agricultural crops and/or animals in some form of spatial arrangement or temporal sequence (ICRAF/World Agroforestry Centre, Nairobi).

b. Agroforestry is a sustainable land management system which increases the overall yield of a land unit and combines the production of crops (including tree crops) and forest plants and/or on the same unit of land and applies management practices that are compatible with the cultural practices of the local population (King and Chandler, 1978).

c. Agroforestry is a land use system that integrates trees, crops and animals in a way that is scientifically sound, ecologically desirable, practically feasible and socially acceptable to farmers (Nair, 1979).

d. Agroforestry is defined as a land use system which integrate trees and shrubs on farmlands and rural landscapes to enhance productivity, profitability, diversity and ecosystem sustainability. It is a dynamic, ecologically based, natural resource management system that, through integration of woody perennials on farms and in the agricultural landscape, diversifies and sustains production and builds social institutions (National Agroforestry Policy, 2014).

Scope and Importance of Agroforestry

Agroforestry has an excellent scope in the context of Indian Agriculture due to its intrinsic relation with traditional agricultural practices. The scope is very significant especially in meeting the domestic (food, fibre, fuel and fodder) and industrial needs (timber, pulp, paper, match, veneer, plywood, packing case, pencil, sports goods, furniture etc.) besides protecting the environment. Agroforestry is the major tool to achieve the envisaged target of forest cover (33%) set in the National Forest Policy 1988.

Role of Agroforestry

a. Agroforestry for meeting fuelwood requirement

b. Agroforestry for meeting the fodder requirements

c. Agroforestry for timber requirements

d. Agroforestry for food security

e. Agroforestry for industrial requirement

f. Agroforestry and climate mitigation

g. Agroforestry and soil Productivity

h. Agroforestry and employment

a. Agroforestry for meeting fuelwood requirement

Agroforestry has wider scope of meeting the domestic requirement of people particularly fuel and fodder needs. The fuelwood requirement of the country is ever increasing which was estimated around 200 million tonnes in 2011, and has now increased to over 240 tonnes. Most of these wood requirements are met predominantly from agroforestry which can bring down the pressure on natural forests of the country. It is also estimated that 54% of the fuelwood requirements of the country is met through agroforestry.

b. Agroforestry for meeting the fodder requirements

Trees and shrubs contribute substantial amount of leaf fodder especially during the lean period. Promotion of silvipasture systems can help to meet the increasing demand for fodder requirement. Incorporation of fodder trees species like *Acacia*, Subabul, *Dalbergia*, *Hardwickia*, *Holoptelia*, *Pithecolobium*, *Morus*, *Moringa*, *Melia* etc. will extend greater scope in meeting the fodder requirements of the country.

c. Agroforestry for timber requirements

Agroforestry plays a key role in meeting the timber requirement of society. The growing population and associated socio-economic development require huge volume of timber. It is estimated that around 31 million m^3 of wood is obtained from farm forestry and trees outside forest to meat the timber requirement also. It is estimated that almost 50% of the wood requirement comes from non-forest source particularly agroforestry and farm forestry.

d. Agroforestry for food security

Food production in India has increased four folds during the last six decades, however the improvement made in food supply have been insufficient to fulfill the nutritional needs of the poor people in the country. Increasing the production and consumption of protein rich food will be necessary to solve this issue. Agroforestry by virtue of diversity integrates pulses, fruits, leafy vegetables and other perennial food crops offers greater scope in ensuring the food security. It is estimated that food production in the country will be around 320 million tonnes and Agroforestry is valued to contribute about 25 million tonnes per annum.

e. Agroforestry for industrial requirement

Agroforestry systems are capable of providing wide range of raw materials to several wood based industries. In India, there are around 26,000 wood based industries which are operating at around 40% operational capacity for want of sustained supply of industrial wood. The wood based industries in the country particularly pulp and paper, ply and veneer, match and splints, sports goods, pencils, furniture etc. have grown substantially both in production and installed capacity. Due to policy and legal regulations, the industries which were once deriving raw materials from the natural forests are forced to meet their raw material requirement by themselves. Hence, National Forest Policy (1988) emphasized and directed wood based industries to raise their own captive plantations, promote farm forestry and agroforestry initiatives. Hence, agroforestry has a greater scope in meeting industrial raw material requirement because the country is importing industrial wood to the tune of around 8 billion USD which when replaced by agroforestry could auger well for improving the livelihood of the farmers besides reducing the pressure on the exchequer.

f. Agroforestry and climate mitigation

Increasing human activities have contributed to increase in CO_2 concentration to the level of 396 ppm and associated other Green House Gas (GHG) emissions which in turn have resulted in drastic changes in climate patterns around the globe. These changes in climate pose a serious implication on global warming resulting in melting of Polar regions, rising of sea levesl, changes in rainfall pattern and consequently their implications on Agriculture. Generally, trees have a huge potential in mitigating these issues by acting as a sink for most prevalent GHGs. The role of trees as a major carbon sink is scientifically acknowledged as the most important means for successfully immobilizing large quantities of carbon for long years. It is estimated that for each and every one ton of wood produced, 0.5 tonnes of carbon is sequestered. Global studies estimate that agroforestry has the potential to be the major source for effective climate mitigation especially in the tropical zone of the globe.

g. Agroforestry and soil productivity

Soil is one of the critical factors for agricultural productivity and in India, over 175 million ha of land area is degraded as a result of soil erosion. It is believed that to produce 1cm soil layer, it takes over 600 years. But 1 cm soil layer can be easily washed by a simple rain in few seconds. Trees by virtue of their vast root system help to protect the soil and arrest the erosion. The tree canopy reduces the velocity of rains and help to infiltrate into the soil. Incorporation of Nitrogen fixing trees in Agroforestry help to influence soil fertility positively due to Nitrogen (N_2) fixation.

h. Agroforestry and employment

Agroforestry activities provide huge volume of employment and income generation activities. It is estimated that Agroforestry activities like land preparation, nursery, pitting, planting, management, felling and transportation could generate 200-500 mandays of employment per hectare depending on the tree species and type of agroforestry systems.

Classification of Agroforestry systems

Agroforestry has been classified into the following types.

S.No	Basis of classification		Various Agroforestry systems
1.	Structural basis	Nature and mixture of crops	(a) Agri-Silviculture (b) Silvipasture (c) Agri-Silvipasture (d) Miscellaneous system
		Arrangement of different components	(a) Spatial arrangements (b) Temporal arrangements
2.	Functional basis		(a) Productive agroforestry (b) Protective agroforestry
3.	Socio- economic basis		(a) Commercial Agroforestry (b) Subsistence Agroforestry (c) Intermediate Agroforestry
4.	Ecological basis		(a) Humid and Sub-humid Agroforestry (b) Arid and Semi-arid (c) Highland agroforestry

(I) STRUCTURAL BASIS

Based on nature of components

(a) Agri-Silviculture

In this system, agricultural crops are grown in combination with tree crops. The following 9 subsystems are widely practised in India.

S.No.	Subsystem	Common tree-crop combination
1	Improved fallows	This system is practised in North Eastern India where existing shifting cultivation is improved through inclusion of trees with high yielding crop varieties
2	Taungya system	It is called hill cultivation and common in West Bengal and Uttar Pradesh. In this method forest, plantations are established and agricultural Crops are grown till the tree canopy closes.
3	Multiple tree gardens	This method is common in all parts of country where tree crops grown in association with fruit crops, fruit, and fodder and woody timbers. The major fruit trees included are Mango, Jack along with timber crops like *Acacia*, *Albizia*, Teak etc.

S.No.	Subsystem	Common tree-crop combination
4	Alley cropping	It is otherwise called as hedge-row intercropping. In this method, perennial tree crops are grown as hedges and agricultural crops are grown in alleys. The hedge species are predominantly N_2 fixing and leaves are collected and used for mulching which improves soil fertility. The common tree crops are 1. *Leuceana* *2. Gliricidia* 3. *Sesbania grandiflora* 4. *Cassia siamea* Agricultural crops like cereals, pulses and oil seeds are grown besides green manure crops.
5	Trees with plantation crops	This method is practised mostly in plantations like Tea and Coffee in South India and North East India. Tree crops like *Cassia siamea*, *Grivellia robusta* are grown as shade trees for tea and coffee.
6	Multi-purpose trees and shrubs on farm land	In this system, multipurpose tree species are planted sporadically on bunds or on the boundary or within the field. The main objective is to produce various products and protect the ecosystem. Eg: *Azadirachta, Leuceana, Acacia, Casuarina, Albizia* etc.
7	Home shed plantations	This system is adapted extensively in high rainfall regions of North Eastern states and Kerala. The crop combination includes Jack, Mango, Coconut, Arecanut, Nutmeg, Pepper, Vegetables, grasses etc. It is also called multi–tier agroforestry. Judicious incorporation of various components like herbaceous along the ground,lower shrubs in the first storey followed by fruit and fodder trees along with spice crops in the middle layer and timber and plantation trees in the top layer is usually practised in this system.
8	Shelterbelts	It is established in areas of heavy wind velocity and this system consists of a block of trees consisting of several rows of trees established at right angles to the prevailing winds. Shelterbelts are established upto 50 m width with typical trianglular shape. Choice of species – **Grasses:** *Saccharum, Panicum and Cenchrus* **Shrubs :** *Calotropis, Cassia, Dodonea* spp. **Trees:** *Acacia, Dalbergia, Lannea, Eucalyptus, Casuarina, Bamboos*
9	Wind breaks	Wind breaks are strips of trees planted to protect hills, homes, canals, agricultural fields and other areas from wind. Preferred species : *Casuarina, Eucalyptus, Cassia and Dalbergia*

(b) Silvipastoral systems

It is one of the major agroforestry systems wherein woody tree species are grown along with grasses. The tree species incorporated in silvipastoral system may be of multiple utility from fodder to timber. The silvipastoral system is further classified into three categories and they are

(i) Trees and shrubs as pasture

(ii) Protein based fodder banks

(iii) Fodder trees, shrubs and hedges as live fence

(i) Trees and shrubs as pasture

It is one of the silvipastoral systems in which multipurpose trees and shrubs are planted sporadically with or without systematic pattern in the pasture lands. The main idea of introducing trees on pasture land is to supplement fodder availability during the lean season.

Preferred species: *Acacias, Prosopis, Tamarind, Emblica, Holoptelia, Hardwickia*, Neem.

(ii) Protein based fodder banks

It is one of the important silvipastoral systems in which protein rich tree species are planted in and around pasture lands. These tree species provide protein rich leaf fodder to supplement fodder production during the lean season.

Protein rich tree species are

- *Acacia* spp.
- *Albizia*
- *Leuceana*
- *Gliricidia*
- *Sesbania*
- *Prosopis*

(iii) Fodder trees, shrubs and hedges as live fence

In this system, several fodder trees and hedges are planted around the farm land as live fence to protect the crops within the farm land from all biotic influences. The preferred species for live fences are

- *Gliricidia sepium*
- *Sesbania grandiflora*
- *Acacia leucophloea*
- *Leuceana leucocephala*

- *Protium cordatum*
- *Euphorbia*

(c) Agrisilvipastoral system

The system is the hybrid between Silvipastoral and agrisilvicultural system. Here, the land is managed to get agricultural crops, forest trees and livestock. Home garden is one of the classical examples for agrisilvipstural system which integrates animal rearing with crop and forest tree production. Integrated mixing of trees, annual crops and livestock for various purposes like soil and water conservation, green manure production etc. also represents agri silvipastoral system.

(d) Miscellaneous System

There are several agroforestry practices that exist in India which integrates tree crops with honey bee, fish culture and multipurpose mixed wood lots. They are classified as

(i) Silvi-apiary

(ii) Aquaforestry

(iii) Mixed woodlots

(i) Silviapiary

It is one of Agroforestry systems which integrates trees with honey production. In this system, profusely flowering trees are planted in the farm lands as boundary plantation or sporadically to attract honey bees to collect nectar from the flowering trees.

Eg: *Azadirachta indica, Syzygium cuminii, Gliricidia sepium, Cassia* spp. etc.

(ii) Aquaforestry

This method integrates trees with fish production. The leaves and pods of many trees act as a feeding material for fish. Hence, tree species are planted all around the fish ponds which will ensure fish production besides stabilizing the bunds of fish ponds.

Preferred species are

- *Leuceana leucocephala*
- *Morus alba*
- *Acacia leucophloea*

(iii) Mixed woodlots

Under this system, many multipurpose tree species are grown mixed for various purposes of wood production, soil conservation, food production, soil reclamation.

Eg. 1. Teak – (Border) + *Melia* (Plywood) + *Casuarina* (Pole & Paper), 2. *Melia* + *Gmelina* + Teak

Arrangement of components

Agroforestry system is classified based on components arrangements into two types

a. Spatial arrangements

In this case, agroforestry systems are classified based on species mixture. It may be densely mixed crop as practised in home gardens or sparsely mixed crops as in pasture lands.

b. Temporal arrangements

This system takes into consideration the arrangement of crops in the agroforestry systems. The temporal arrangements may be coincidence like coffee under shade trees or concomitant (Agrisilvicultural systems), Intermittent (Hedgerow inter-cropping) and Interpolated (Hedge gardens).

(II) FUNCTIONAL BASIS

Agroforestry is classified into two types bases on its functional roles.

a. Productive Agroforestry

In this system, agroforestry is designed only for its productive function like producing wood, food, fuel, fodder, fibre, medicinal products, industrial raw materials etc.

Eg. - Teak – Timber

Eucalyptus – Pulpwood

Casuarina – Pole and Paper

Ailanthus – Matchwood

Melia – Plywood, Fodder

Subabul – Fodder & Paper

b. Protective Agroforestry

In this system, the tree species are incorporated in farm lands mainly for its protective role like windbreaks, shelterbelts, soil and water conservation, soil reclamation, shade requirements etc.

Eg: *Casuarina, Eucalyptus, Dalbergia – Windbreak*

Neem, *Pungam* – Shade

Acacia – Soil improvement and soil reclamation.

(iii) SOCIO-ECONOMIC CLASSIFICATION

Agroforestry is classified into three categories based on the socio-economic criteria.

a. Commercial Agroforestry
b. Subsistence Agroforestry
c. Intermediate Agroforestry

(a) Commercial Agroforestry

It is one of the most popular agroforestry systems grown exclusive for commercial purpose. Industrial forestry, farm forestry, plantation forestry come under this category. The main objective is to produce, the commercial wood and other products like timber, paper, match, plywood, tannin or resin production.

(b) Subsistence Agroforestry

In this system, agroforestry is practised towards satisfying the basic needs of farmers and it is solely by the farmer and his family members.

(c) Intermediate Agroforestry

It is hybrid between commercial and subsistence agroforestry wherein production of perennial crops and subsistence crops are undertaken by medium to small sized farmers. In this system, the cash crop satisfies the economic needs and the food crops satisfy the subsistence needs. Many agroforestry systems practised in India represent this category.

(iv) ECOLOGICAL CLASSIFICATION

Ecologically Agroforestry systems are classified into several types but the most prominent ones are

a. Humid and Sub-humid Agroforestry
b. Arid and Semi-arid
c. Highland agroforestry

(a) Humid and Sub-humid Agroforestry

It is one of the agroforestry systems practised in humid and sub humid systems. This system is based on the local needs, nature of components and its major functional characteristics. The choice of Agroforestry system depends on social and economical practices.

Preferred species

1. *Acacia mangium*

2. *Albizia falcataria*
3. *Albizia lebbeck*
4. *Albizia procera*
5. *Gliricidia sepium*
6. *Leucaena leucocephala*
7. *Sesbania glandiflora*
8. *Calliandra calothyrsus*

(b) Arid and Semi-Arid Agroforestry

The most predominant ecological classification of agroforestry is arid and semi-arid practices which represents majority of such practices in India.

The preferred species

- *Acacia albida*
- *Acacia nilotica*
- *Acacia tortilis*
- *Albizia lebbeck*
- *Albizia amara*
- *Leucaena leucocephala*
- *Prosopis cineraria*
- *Sesbania aegyptica*
- *Hardwickia binnata*
- *Dalbergis sissoo*
- *Melia azadirach*
- *Azadirachta indica*

(c) Highland Agroforestry

This is one of the agroforestry systems practised mostly in hilly areas. The preferred species are

- *Acacia mearnsii*
- *Leuceana diversiflolia*
- *Albizia falcataria*

TREES SUITABLE FOR AGROFORESTRY

Agroforestry requires trees species with special characteristics and they are

- The trees should be a fast grower and comenable for easy management.

- Trees species should have ability to fix atmospheric nitrogen.
- The tree species should not compete with agriculture crops for moisture, nutrients, water, space etc.
- The tree species should have high survival rate and should have easy establishment.
- The tree species should not produce any toxic substance and should not cause alleplopathic effect.
- The tree species has wider adaptability and acceptability without any controversy.
- The tree species should have high productivity, higher yield and maximum profit to the growers.
- The tree species should have excellent coppicing ability and also amenable for pruning and pollarding.
- The trees should have multipurpose utility like timber, fodder, fuel, nitrogen fixer etc.

Multipurpose tree species

The following multiple utility characters are considered essential for deploying trees in Agroforestry systems.

- The trees should have high calorific value and should also be used as fuelwood and for energy generation.
- The trees species should be able to produce timber and also amenable for poles and post in scaffolding industries.
- The trees should be suitable as raw materials for various wood based industries like pulpwood, paper, plywood, matchwood etc.
- Trees should be able to produce flowers and fruits for both animal and human consumption.
- The leaves should be palatable for human and animal consumption.
- The tree should act as a source for honey, silk and lac insects.
- The bark should contain tannin usable in leather industry.
- The trees should be able to produce gums, resins, dyes etc.
- The trees should give oils amenable for both domestic and human utility. The multipurpose trees species identified need not have all the properties mentioned above but should contain at least two or three characters which will facilitate its incorporation in agroforestry systems.

Benefit of Agroforestry

(i) Environmental benefits

- Agroforestry reduces the pressure on forest and enhanced better protection of ecological systems.
- It augments biological diversity by favouring environmental conditions.
- It improves microclimate of the area by lowering of soil temperature, reduction of evaporation and maintenance of soil moisture.
- It reduces run-off, nutrient leaching and soil erosion.
- It improves the soil fertility through addition of litter-fall.
- It enhances efficient recycling of nutrients with in soil at different depth.
- It improves soil structure by the addition of organic matter of soil.

(ii) Economic benefits

- Agroforestry yields increased outputs per unit area (more crop per unit area).
- Agroforestry system reduces the total crop failure.
- All these benefits resulted in increase of farm income

(iii) Social benefits

- Agroforestry improves the rural living standards by increasing their income in a sustained manner.
- It improvement the health and nutrition of the peoples by providing diverse, nutrient rich farm produce.
- Multiple outputs without the risks of total failure stabilize the farm communities.

Constraints in agroforestry

- Depression in crop yields due to interference effects caused by the tree
- Delayed liquidation of planting investments due to long gestation period
- Increased damage to crops due to birds which the tree attract
- Increased damage to crops due to pests for which the tree serve as alternate hosts
- Allelopathy

TREE CROP INTERACTION

Agroforestry is a wise integrated model with 2 or more farm enterprise such as Agriculture, Horticulture, Forestry and livestock in same unit of land. Once these components co-exist, there arises the interaction because all members of the agroforestry system utilize the same reserves of growth resources such s light, nutrients, water and Co_2.

Each and every component influences other component of the agroforestry system. The net effect of component may be either positive or negative. The success of agroforestry relies heavily on the wise exploitation of various positive interactions and minimizing or totally eliminating negative interactions. Various interactions that exist in agroforestry are discussed below.

The major yield reducing impacts of tree crop interactions arise from competition for light, water, nutrients and allelopathy.

(i) Competition for light

The amount of light and its intensity is an important factor that decides the photosynthetic activity of plants. Studies also show that particular amount, intensity and length of light (photoperiodism) is much needed for most of the crops to reproduce (flower and fruit) which is ultimately our yield in agricultural crops trees with branching crown always cast their shades over the inter-crops and reduces the yield apparently. During the 1st two years the trees won't have dense crown and allows maximum sunlight. The competition likely to be severe as the year's progress.

(ii) Competition for Moisture

The prime most assumption of agroforestry system is that the trees have deep roots to take moisture and hence do not compete with the shallow rooted crops. Many of the agroforestry trees were reported to have lateral roots dominating the top 30cm of soil which comes into competition with the inter-crops. this competition is further associated with high root density, high root shoot ratio, high root length per unit soil volume.

(iii) Competition for nutrients

The merit of an agroforestry system is assessed based on the yield of agricultural crop. The intercrop being the smallest component of the system has small root system hardly reaching not more than 50 cm depth. Hence it can only feed on the top most soil layer. The tree with large root system has the advantage of utilizing the top soil layer and can reach and exploit soil volume beyond the reach of crop component. This situation makes the herbaceous crop to be severely affected and deprived of nutrients which further drastically affects yield in crops.

(iv) Allelopathy

The term Allelopathy was first coined by Hans Molisch which has its origin from Greek meaning "mutual harm" or "suffering".

Allelopathy is a common biological phenomenon by which one organism produces biochemicals that influence the growth, survival, development and reproduction of other organisms. The inhibitory chemicals known as allelochemicals are mostly a sub-set of secondary metabolites present in any parts of the plant viz., leaves, bark, root, twigs, flowers and fruits. these chemicals are released into the environment through leaching, volatilization, exudation and decomposition. the inhibitory role of Acacias, Casuarinas and Eucalyptus on germination, growth and development of various crops have been well chronicled.

Management stratergies for above ground interaction

(i) Tree management / crown management

There are certain management techniques which are applied to trees and shrubs in agroforestry systems. Some of these techniques are similar to those used in the management of trees in forestry plantations, but others are different. The most important management techniques with regard to the part of the tree which is above ground are:

a. Pruning
b. Lopping
c. Pollarding
d. Coppicing
e. Thinning

a. Pruning

Removal of branches from the lower part of the tree crown is known as pruning or side pruning. While pruning a tree, branches are always cut near the stem.

The objectives of pruning in agroforestry are three fold:

- Reduction of shade for crops near the tree
- Improving the quality of the trunk, mainly for timber and poles
- Early harvest of branch wood for fuel or other use.

Pruning should be done at least up to the height the adjacent crops if trees are growing in fields. Such pruning facilitates farming operations and reduces competition. The best time for pruning is towards the end of the dry season when the work will not interfere with growing crops and when the workload in other agriculture tasks is not so heavy.

b. Lopping

Lopping is distinguished from pruning in that branches are not cut from the base. Also lopping is not always done starting from the lower part of the tree but can be more haphazard. If any selection of branches is made, the main criterion is often a good green leafy biomass since the lopping is usually done to obtain branches for fodder.

c. Pollarding

If all the branches and the top part of a tree are cut off this is known as pollarding. There can be several objectives with pollarding:

- Early harvest of wood, fodder or other biomass
- Production of wood or fodder that is out of the reach of livestock, hence there is no need for protection from browsing
- Reduction of shade for crops near by
- Regeneration of the tree crown to promote growth of the trunk for timber or poles

d. Coppicing

Systematic coppicing is applied as the management technique in alley cropping, and it may be an option for trees on soil-conservation structures. In such a situation coppicing may be done annually, but in other situations, e.g. regeneration of *Eucalyptus* for pole production, it may be much less frequent. In that case, an interval of 6-8 years may be more suitable.

e. Thinning

Trees established by direct seeding or that have been planted with little space between them will soon start to compete with each other. A dense stand initially promotes straight growth and small branches, but later the trees must be thinned otherwise they will grow too slender and eventually not reach the desired size. Thinning is particularly important for trees grown in woodlots, but applies also to other situations where trees are growing close to each other. Thinning can, for example, be done by removing every second tree or two out of every three trees. Thinning is also a way of obtaining some early harvest.

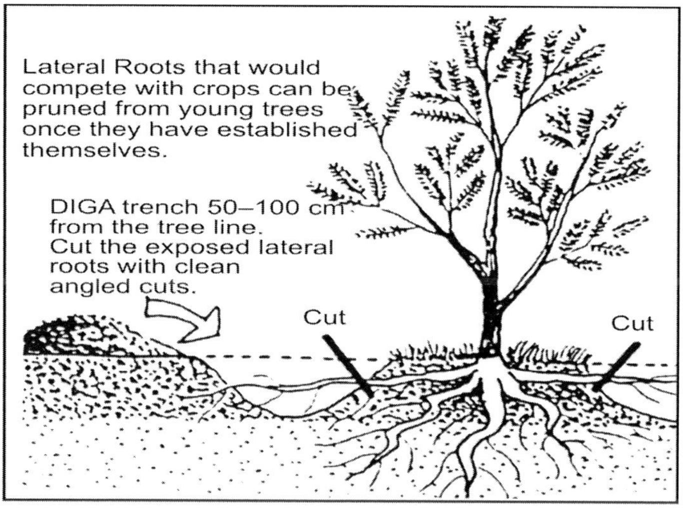

Management strategies for below ground interaction

Root management and root pruning

Trees growing in cropland can have their shallow roots cut 0.3-0.6 m from the trunk when they reach a height of 2-3 m. This is applicable to species which would otherwise compete with crops. This also serves to minimize competition.

There is an inherent conflict in agroforestry between expected favorable effects of tree root systems, e.g. on soil fertility and nutrient cycling, and competition between tree and crop roots.

Root management attempts to optimize root functions and to stimulate facilitative and complementary interactions. It makes use of the plasticity of root systems to respond to environmental factors, including other root systems, with altered growth and physiology. Root management tools include species selection, spacing, nutrient distribution, and shoot pruning, among others.

Root distribution determines potential zones of root interactions in the soil, but are also a result of such interactions. Plants tend to avoid excessive root competition both at the root system level and at the single-root level by spatial segregation. As a consequence, associated plant species develop vertically stratified root systems under certain conditions, leading to complementarity in the use of soil resources.

Parameters of root competitiveness, such as root length density, mycorrhization and flexibility in response to water and nutrient patches in the soil, have to be considered for predicting the outcome of interspecific root interactions. The patterns of root activity around individual plants differ between species; knowing these may help to avoid excessive competition and unproductive nutrient losses in agroforestry systems through suitable spacing and fertilizer placement.

NUTRIENT CYCLING

Nutrient cycling is the continuous process of transfer of minerals in and out among various nutrient pools. Nutrient cycling is very significant to maintain the soil fertility. The following are the major nutrient cycle prevalent in Agroforestry systems.

i. Carbon cycle

ii. Nitrogen cycle

iii. Oxygen cycle

iv. Phosphorus cycle

Nutrients are elements or compounds that are essential for the growth and survival of plants. Plants require large amounts of nutrients such as nitrogen (N), phosphorus (P), carbon (C), hydrogen (H), oxygen (O), potassium (K), calcium (Ca) and magnesium (Mg), but only small amounts of others such as boron (B), manganese (Mn), iron (Fe), copper (Cu), zinc (Zn) and chlorine (Cl) (micronutrients).

Forest nutrient cycling is defined as the exchange of elements between the living and nonliving components of an ecosystem. The processes of the forest nutrient cycle include:

- Nutrient uptake and
- Storage in vegetation perennial tissues,
- Litter production,
- Litter decomposition,

Nutrient transformations by soil fauna and flora, nutrient inputs from the atmosphere and the weathering of primary minerals, and nutrient export from the soil by leaching and gaseous transfer.

Features of the major nutrient cycles

Element	Uptake by the trees	Tree uptake	Limiting situation
Carbon	Atmosphere	Atmosphere	Atmospheric concentration may limit growth
Oxygen	Atmosphere	Atmosphere	Waterlogged soils
Hydrogen	Atmosphere	Atmosphere	Extremely acidic and alkaline conditions
Nitrogen	Soluble $N0_3$ and NH_4 N_2 for nitrogen fixing species	Soil organic matter; atmospheric N_2 for nitrogen fixing species	Most temperate forests, many boreal forests and some tropical forests
Phosphorus	Soluble phosphorus	Soil organic matter; adsorbed phosphate and mineral phosphorus	Old soils high in iron and aluminum, common in subtropical and tropical environment
Potassium calcium Magnesium	Soluble Resoluble Ca^{2+}/ soluble $Mg^{''+}$	Soil organic matter; exchange complex and minerals	Miscellaneous situations and some old soils

Gersmehl's Nutrient Cycles

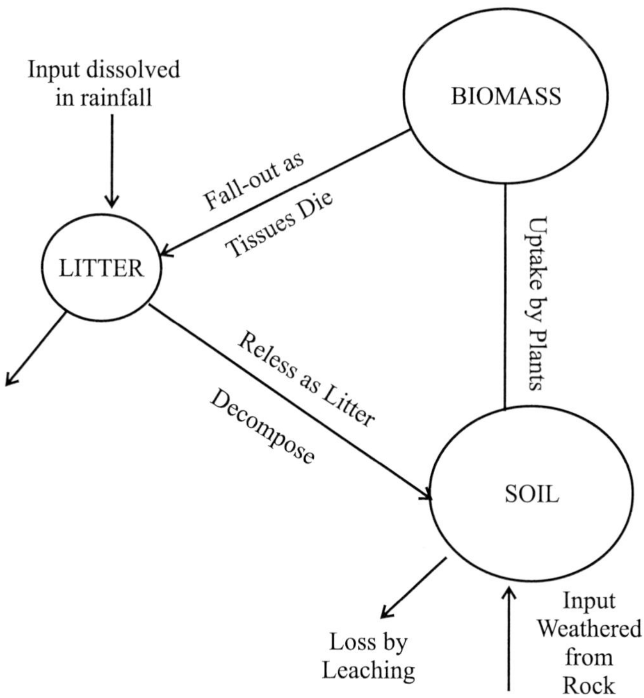

Nutrient cycling in forest ecosystems is controlled primarily by three key factors: climate, site, abiotic properties (topography, parent material), and biotic communities.

Nutrient cycling in agroforestry system

In agroforestry systems, trees can provide nitrogen inputs by 2 processes biological nitrogen fixation in the case of nitrogen-fixing tree species(NFTS),and deep nutrient capture. Although the magnitude of biological nitrogen fixation is methodologically difficult to quantify, the overall recorded rates of fixation for trees both leguminous as well as non-leguminous range from20to 300 kg N/ ha /yr. The ad vocation of inclusion of NFTS in agroforestry systems has led to the belief that nitrogen fixed in the root nodules may be used by the companion crop though direct evidences to support it are still lacking. NFTS can also supply considerable nitrogen to crops through litter fall. On the other hand, there is ample evidence that trees such as cherry and mandarin, where nitrogen-fixation does not occur, accumulate as much or more N in their leaves as do NFTS, presumably because of their greater root volume and ability to capture nutrients from sub surface soil. However, these non NFTS are only cycling nutrients and not adding them to the system.

Deep nutrient capture the second process through which trees can provide nitrogen inputs is the uptake of nutrients by tree roots at depths where crop roots are not present. This is an additional nutrient input in agroforestry systems because such nutrients are leached and lost for crop use, but they become an input on being transferred to the soil through tree litter decomposition.

AGROFORESTRY SYSTEMS FOR DIFFERENT ECOLOGICAL ZONES

There are five agroecologial zones in India and they are

i. Himalayan region

ii. Indogangetic Plains

iii. Arid and semi arid Region

iv. Humid and Sub Humid Region

v. Coastal and Iceland Region

i. Himalayan region

AFS	Examples
Trees on Agricultural field or on field bunds	*Grewia optiva* and other trees in Jammu & Kashmir, HP, Sikkim, Other north eastern hill region
Inter cropping on fruit orchards	Orange, other citrus and guava intercropped with cereals, tubers and rhizomes crop on Sikkim, Meghalaya

Plantationcrops under shade of trees	Large cardamom, Coffee,etc under alder in Sikkim and Nagaland; Tea under leguminous trees in Assam and WB
Silvipastoralpractices	Tree leaf fodder from Grewia spp., *Celtis australis* and other species in Himalayas

ii. Indogangetic Plains

AFS	Examples
Trees for soil reclamation (Sodic/saline/degraded soil)	*Prosopsis chilensis, Acacia nilotica, Parkinsonia aculeata* and other for problematic soils.
Fodder trees in degraded grazing lands	*Albizia lebbeck, Bauhnia purpurea, D. sissoo* etc. in grazing lands
Treeson boundaries of Agrifields	*Populus* spp, Tamarindus, *Ceiba pentandra, Eucalyptus* spp etc.
Fodder banks and wood lots	Fodder trees – *Azadirachta indica*, Melia, Albizia spp, *Syzygium cumini* etc. Wood lots – *Casuarina equisetifolia*, Eucalyptus, Bamboos, *Dalbergia sissoo* etc

iii. Arid and semi arid Region

AFS	Examples
MPTson Agricultural land	*Prosopis cineraria, Zizipus* spp, *Acacia senegal, Ailanthus excelsa* etc
Trees for reclamation of degraded soils	*Albizia lebbeck, Casuarina equisetifolia, Acacia tortolis, neem* and other spp.
Windbreaks and shelterbelts	*Acacia nilotica, cassia spp, Azadirachta indica, Tamarix spp, Prosopis chilensis etc.*
Fodder banks	*Prosopis cineraria, Ziziphus mammularia, Salvadora persica etc*
Trees on rangelands (Silvopastoral system)	*Prosopis chilensis, Acacia senegal, Acacia senegal, Prosopis cineraria, Zizipus spp.* etc

iv. Humid and Sub Humid Region

AFS	Examples
Homegardens/Homesteads	Homegardensin Kerala, Assam, WB etc., have a mixture of trees, shrubs, herbs etc. Coconut, Areca nut, Erythrina, Gliricidia, with black pepper, cocao, cassava, and other cash crops.

Multi type system or plantation crops combination	Coconut, Arecanut, Erythrina and other trees with coffee/banana, pineapple papaya cacao/coffee and black pepper /betel vine/large cardamom.
Multipurpose trees in Agrifields	*Acacia ariculiformis, Aegle marmelos, Albizzia spp, Anogeiscus latifollia, Anthocephalus chinensis, Artocarpus spp etc.*

v. Coastal and Iceland Region

AFS	Examples
Plantation crop combination or multistoried	Coffee under *Erythrina lithosperma*, Cacao with coconut, Black pepper on Gliricidia spp, *Grevellia robusta,* Cardamum under *Toona ciliata,* Artocarpus etc
Trees with Aquaculture	Composite fish culture in ponds and MPTs in Home steads
Mangroves plantations, Homesteads	*Acanthus ilicifolius, Avicenia officinalis, Azadirachta indica, Acacia senegal* etc.
Shelterbelts and windbreaks	*Casuarina equisetifolia, Azadirachta indica, Prosopis chilensis* etc.
Trees on boundaries of Agrifields	*Acacia auriculiformis, Aegle marmelos, Anogeisus latifolia, Calliandra calothyrus* etc

INDUSTRIAL AGROFORESTRY

Definition

Industrial Agroforestry is the specific land use system wherein the industrial utility tree species are planted in an agroforestry system which include both agro and farm forestry.

Scope of Industrial Agroforestry

In India, the demand for wood and wood products are increasing at an alarming rate and it has been estimated that over 152 million m^3 of wood is required during the year 2020 against the actual supply of less than 50 million m^3. This has created over 50% of short fall and necessitated massive promotion of industrial agroforestry. The predominant industries in the country *viz.*, timber, pulp and paper, plywood and veneer, match splints and boxes, pencil, packing, dendro power, biofuel and construction industries are requiring huge volume of wood and hence the industrial agroforestry has an excellent scope of application in India.

Policy directions

The National Forest Policy (1988) has directed all the wood based industries in the country to generate their own raw material needs without depending on forest department supply. It has also guided the industries to promote agroforestry plantations by establishing suitable linkages with farmers by facilitating quality seedling supply, technological support, support price for farm grown wood and also credit and insurance facility. This has stimulated several industries to initiate industrial agroforestry plantations in the country. Similarly, the National Agroforestry Policy (2014) has intensified the promotion of industrial agroforestry plantations by establishing suitable linkages with all stakeholders. It has also directed the government to liberalize the timber transit rules in order to attract more farmers towards industrial agroforestry. The policy also directed to extend credit and insurance facilities to agroforestry plantations besides establishing necessary market mechanism.

Major Wood based industries and the prioritized species

S.No.	Major industry	Species promoted and deployed
1.	Timber and construction industry	*Tectona grandis, Dalbergia latifolia, Gmelina arborea, Shorea robusta, Pterocarpus marsupium*
2.	Pulp and Paper	*Eucalyptus tereticornis, Eucalyptus camaldulensis, Casuarina equisetifolia, Casuarina junghuhniana, Acacia hybrids, Acacia auriculiformis, Acacia mangium, Leucaena leucocephala, Bamboos* etc.
3.	Plywood	*Populus deltoides, Eucalyptus, Melia dubia, Neolamarckia cadamba, Toona ciliata, Acrocarpus fraxinifolius* etc.
4.	Matchwood	*Ailanthus excelsa, Populus deltoides, Albizia falcataria*
5.	Dendro Biomass	*Prosopis, Leucaena leucocephala, Dalbergia sissoo, Acacias, Casuarina, Eucalyptus*
6.	Biofuels	*Jatropha curcas, Pongamia pinnata,Callophyllum inophyllum, Bascia latifolia* etc.

Successful industrial agroforestry plantations

Poplar based industrial agroforestry plantations in North India promoted by WIMCO and Eucalyptus by ITC in South India are successful evidences of industrial agroforestry plantations. The model developed in Tamil Nadu through Consortium of Industrial Agroforestry conceived by Tamil Nadu Agricultural University involving pulp and paper (TNPL and SPB), plywood (Century ply / Sharon/Ambiply), timber (Suresh timbers), Biofuel (Auromira Energy and Bannari Amman Group of Companies), Match (Vasan Matches and Ideal splints) have resulted in establishment of over 70000 ha of exclusive industrial agroforestry plantations. These developments have ensured availability of quality planting material, technology support, felling and transportation besides assuring support price for farm grown industrial wood trees. The introduction of tree insurance scheme in Tamil has attracted more farmers towards industrial agrofrestry.

Agroforestry in general and industrial agroforestry in particular is gaining significant attention across the world towards meeting the domestic and industrial wood requirement besides catering the needs of the risks associated with climate change. Though the agroforestry is an age old practice, still the achievement in this front is dismally modest for want of suitable science and technological advancements coupled with the associated supply chain process. To resolve these issues, several wood based industries are involved in promotion of industrial agroforestry through technological, organizational and marketing interventions which are at various stages of development and implementation.

NATIONAL AGROFORESTRY POLICY, 2014

Need for Agroforestry Policy in India

- Absence of a dedicated and focused national policy and a suitable institutional mechanism
- Lack of an integrated farming systems approach
- Restrictive regulatory regime
- Inadequate attempts at liberalization of restrictive regulations
- Insufficient research, extension and capacity building
- Dearth of quality planting material
- Institutional finance and insurance coverage
- Weak market access for agroforestry produce
- Industry operations at a sub-optimal level

Goal

The major policy goals are:

- Setting up a National Agroforestry Mission or an Agroforestry Board to implement the National Policy by bringing coordination, convergence and synergy among various elements of agroforestry scattered in various existing, missions, programmes, schemes and agencies pertaining to agriculture, environment, forestry, and rural development sectors of the Government.

- Improving the productivity, employment, income and livelihood opportunities of rural households, especially of the smallholder farmers through agroforestry.
- Meeting the ever increasing demand of timber, food, fuel, fodder, fertilizer, fibre, and other agroforestry products; conserving the natural resources and forest; protecting the environment and providing environmental security; and increasing the forest/tree cover, there is a need to increase the availability of these from outside the natural forests.

Basic Objectives

The basic objectives of the National Agroforestry Policy are to:

- Encourage and expand tree plantation in complementarity and integrated manner with crops and livestock to improve productivity, employment, income and livelihoods of rural households, especially the small holder farmers.
- Protect and stabilize ecosystems, and promote resilient cropping and farming systems to minimize the risk during extreme climatic events.
- Meet the raw material requirements of wood based industries and reduce import of wood and wood products to save foreign exchange.
- Supplement the availability of agroforestry products (AFPs), such as the fuel-wood, fodder, non-timber forest produce and small timber of the rural and tribal populations, thereby reducing the pressure on existing forests.
- Complement achieving the target of increasing forest/tree cover to promote ecological stability, especially in the vulnerable regions.
- Develop capacity and strengthen research in agroforestry and create a massive people's movement for achieving these objectives and to minimize pressure on existing forests.

Strategy

- Establishment of Institutional Setup at National level to promote Agroforestry
- Simple regulatory mechanism
- Development of a sound database & information system
- Investing in research, extension and capacity building and related services
- Improving famers' access to quality planting material
- Providing institutional credit and insurance cover for agroforestry

- Facilitating increased participation of industries dealing with agroforestry produce
- Strengthening farmer access to markets for tree products.
- Incentives to farmers for adopting agroforestry
- Promoting sustainable agroforestry for renewable biomass based energy

Pathway for achieving Policy Deliverables

- Mainstreaming agroforestry in agriculture policies and strategies
- Enabling environment and legislation and simplify regulations related to forestry
- Trees species for the economic and ecological benefits of the farming community.
- Public Private Partnership (PPP) to be encouraged for economic returns and contributing ecological services.
- Providing quality and certified planting material.
- Data collection with source of agroforestry produces at National level.

■■■■

CHAPTER - 11

SOCIAL FORESTRY

The promotion of forestry in the country has taken different dimensions and one of the significant dimensions is social forestry and its promotion. The concept of social forestry though is an age old, but gained scientific attention after the pronunciation of the word social forestry by Westoby. Since then, social roles of forestry have gained significant attraction in all the forest development activities. The key features of social forestry include development of plantations for the people, by the people and to the people there by ensures participation of the society in each every stage of plantation developmental activities. This chapter has been presented with the following sub headings.

- Definition
- Principles
- Objectives
- Facets of social forestry
- Social forestry programmes in India

Definition

Social Forestry is the forestry for the people, by the people and of the people. Westoby, who first time used the word Social forestry defined it as forestry that aims at producing flow of protection and recreation benefits to the community.

Principles of social forestry

Social forestry has a definite principle of people's participation from the inception upto completion of the project. In regular forestry programmes, trees are planted and managed by Government agencies whereas in social forestry, people's participation is mandatory from planting to utilization. People's participation sometimes is voluntary and many times compulsory participation is ensured.

OBJECTIVES OF SOCIAL FORESTRY

i. Increasing Forest Area and Restoring Ecological Balance

- Moisture conservation—trees take water from the lower soil strata and bring it to the upper layers through long tap root system and, also, trees check evaporation of water;
- Soil conservation—trees help in checking erosion by wind and water;
- Natural habitat conservation—trees provide habitat to many birds and animals, some of which are agro-friendly.
- Improve the environment for protecting agriculture from adverse climatic factors.

ii. Meeting Basic Rural Needs

- Social forestry satisfies the basic rural needs referred to as '5Fs'—food, fuel, fodder, fertiliser (green manure) and fibre. The large-scale depletion of easily accessible forests has resulted in acute scarcity of fuel-wood and fodder. What is disturbing is that the deficit in fuel wood is met by using cow-dung cakes, thus wasting a rich and cheap source of manure.
- Trees also supply the raw material for various small and village industries through small timber and minor forest produce.
- Increase the supply of wood fuel for domestic use, small timber for rural housing, fodder for livestock, and minor forest produce for local industries.

iii. Ensuring Better Land Use

- Social forestry helps achieve a balanced and viable land use by checking soil erosion, facilitating reclamation of marginal lands, checking waterlogging and by bringing about monolithic integration of forestry, agriculture and animal husbandry.
- Increase the natural beauty of the landscape; create recreational forests for the benefit of rural and urban populations.
- To raise the standard of living and quality of life of rural and urban people.

iv. Generation of Employment

- Social forestry operations have the potential of improving the employment situation in rural areas especially during the lean agricultural season. This helps in stabilising incomes of weaker sections of Society.

v. Controlling Pollution

- Trees are known to absorb harmful gases and release oxygen. This way they help reduce air pollution especially in urban areas.

Facets of Social Forestry

Social forestry has different facets and applications and the most pronounced social forestry practices are described below-

I. Farm Forestry

It is a type of social forestry which involves planting of tree species in and around the farm or village lands or community lands. The main objectives of farm forestry are

1. To provide basic needs of society viz., timber, firewood, agricultural implements etc.
2. To provide fodder to the cattle
3. To provide other production functions like fruits, leaves, bark etc.

II. Agroforestry

It is one of the facets of social forestry which primarily aimed at sustainable land management system by incorporating tree crops and / or animal production in same unit of land.

III. Community forestry

Community forestry is a type of social forestry wherein trees are planted by the community for fuel, timer, fodder and fruit. In this community forestry both government and non-government land are utilized.

IV. Urban forestry

It is a kind of social forestry wherein trees are planted in urban areas for the purpose of improving environment and landscape. Flowering and other ornamental trees or shrubs are planted in fallow or urban lands to meet the people's recreational benefits.

V. Environmental forestry

Planting of trees in and around the industrial areas to combat the environmental pollution is called environmental forestry.

Social Forestry Plantations

The social forestry plantation activities have been carried out in India by almost all the states. The successful plantations established through social forestry programme alone are discussed below.

1. Railway Line Plantations

Plantations raised along the railway lines are called railway line plantation or linear plantations.

Objectives

The objectives of planting along railway lines are

i. Stabilization of railway track and protection of railway track against erosion.

ii. Optimum utilization of the land for tree cultivation to produce products needed by the local people.

iii. Checking the illicit encroachment on this land.

iv. Incidental shelterbelt effect of these plantations and

v. Checking of the shifting sand in desert areas getting to the railway tracks.

Orientation

- The safety of the railway traffic, the nature of the land available for planting, the position of the telephone and electric lines, etc. should be considered while planting along the railway lines.
- From the point of view of the safety, the first row of trees should not be planted very close to the railway track.
- The first row of trees is recommended to be planted at a distance of about 7.5 m from the centre of the track.
- Further, the trees in the first row should be such that it may not attain a height more than their distance from the railway track so that in the events of the wind throw the tree tops may not reach the railway line and create traffic hazard.
- The inner side of the curves should not be planted to ensure clear visibility of the track.
- On either side of an unmanned level crossing a length of about 100 m should be left unplanted. No planting should be done under the telephone line while planning orientation of tree lines.

Spacing

- The spacing between the lines and between the tree in a line may be 5 m, 3 m or 2 m depending upon the species to be planted, the fertility of the site and the objects of management.

- Commercial timber species to be harvested at comparatively longer rotations of 30 years or more may be planted at 5m X 5m spacing. Fast growing species like Eucalyptus may be planted at close spacing of 1.5 m X 1.5 m or 2 m X 1.5 m.

Planting

For railway line planting tall seedlings of 3 – 6 ft height are preferred. The pit size should be 60 cm X 60 cm X 45 cm. The pits have to filled with 2 kgs of farm yard manure, 100 grams of neem cake and 50 grams of DAP as basal application. After planting the seedlings have to be irrigated daily till their establishment using tractor mounted tankers. Monsoon planting is preferred for this purpose.

Choice of Species

- The choice of species depends on climatic and site factors and the products required to be produced from such areas.
- Ornamental trees should be planted near the towns, railway stations, road crossings and places of scenic beauty or of tourist interest.
- Trees with brittle stems or branches should not planted especially in the first row. Wind firmness is a desirable character for the species for railway line plantation.
- The plantations along the railway lines are prone to accidental fires originating from engine spark especially in areas infested with tall grass or bush growth The species to be planted in such areas should be able to resists such accidental fires.

Examples: *Acacia, Bauhinia, Casuarina, Dalbergia, Eucalyptus, Pungam, Neem, Peltophorum, Erythrina, Butea, Polyalthia* etc.

2. Canal bank plantations

Plantations raised along canal banks to augment the aesthetic value and stabilizing the canals are called canal bank plantations.

Objectives

The objectives of canal bank plantations are

i. Stabilization of canal banks against erosion.

ii. Checking of shifting sand getting into the canal in desert areas.

iii. Utilization of the available land for tree cultivation and production of tree products, particularly fuel wood and small timber for local people.

iv. Provision of comfort to travelers using canal side road

v. Checking water logging in strips along canals and in adjoining areas.

vi. Shelterbelt effect on adjoining areas.

vii. Improvement of aesthetics in the area.

Orientation

The number of rows to be planted and the position of the first row depend upon the width and nature of the land available for planting along the side of the canal. The first row of trees is generally planted about 7.5 m from the berm of the canal S in the case of multiple row planting. Subsequent rows are spaced at 5m or 3m or 2m depending upon the species to be planted and the objects of management of such canal bank plantations.

Spacing

- The spacing between the trees in a line depend upon the species selected for planning and the desired end product. The first row of trees should preferably be of shade trees.
- Ornamental trees may be planted in the first row or in groves near towns, villages, rest houses, picnic spots, places of tourist interest and crossings of the roads.
- A spacing of 2m x 2m and 3m x 3m are preferred for most species.

Planting

Planting in canal banks may be carried out during non monsoon season or during summer season which will ensure early establishment and the seedlings will be adaptable during monsoon season with excess water flow. Deep planting is preferred for canal bank establishment. A pit size of 45 - 60 cm^3 is preferred.

Choice of Species

- For planting in the first row, the trees with a strong taproot system should be preferred.
- For subsequent lines the species may be selected to provide many tree products needed by the local people. The choice of species also depends on climate, site conditions, object of planting, etc. In places where the plantation strips suffer from water logging, species tolerant of water logging should be selected.

Example : Teak and *Dalbergia sissoo* as in Tanjore canal bank plantation

3. Tank Bed Plantations

Significance of Tank Bed

- Tanks are rainwater collection community irrigation reservoirs.

- The size of tanks range between one and several hundred hectares and are designed to supply irrigation water to a small agricultural community.
- These tanks are usually situated at the edge of a village and have the appearance of a large natural depression partly fiiied with water.
- Most tanks are completely empty during the driest months of the year and full for 15-30 days at the peak of the rainy season so that the whole area of the bed is submerged.
- As water is let out during the dry season to irrigate the surrounding farmland, the level in the tank drops and an increasingly large area of foreshore is exposed.
- The tank foreshore consists of the gradually sloping sides of the tank. In many villages, these areas of temporarily exposed ground are the only substantial areas of common land on which fuel wood plantation might be established.
- Afforestation of tank foreshores is the largest component of the afforestation of communal lands project.
- Most tank foreshores are productive sites because of the availability of moisture. Regular deposition of silt helps to maintain soil fertility.
- The proximity of each tank foreshore to a village is another advantage, both labour and consumer are nearby.
- At the same time, the proximity to village increases the risk of damages by human being and livestock and consequently village co-operation and community involvement are essential for the creation of successful plantations and their proper protection and maintenance. Further, it is a high visibility activity with good demonstration value.

Objectives

i. To reduce the pressure on the forest and produce fuel and fodder for village community.

ii. To convert the wastelands into economically productive lands,

iii. To create financial resources for panchayats to take up welfare measures for community purpose.

iv. To make village panchayats also share holders in community forestry.

Orientation

The species can be planted as block model as an area of tank bed runs from one ha to several ha. It can be treated as irrigated land for all plantation establishment purposes.

Spacing

The spacing adapted for tank bed plantation may vary from 2m × 2m to 3m × 3m. Planting

Planting in tank beds may be carried out during summer season devoid of water. The pit size of 30 cm^3 is preferred along with planting seedlings with 2 - 3 feet height.

Choice of Species

- The main species being used for planting on tank foreshores is *Acacia nilotica.* It has the essential characteristics of being capable of surviving prolonged submersion and periodic flooding.
- It grows quickly and produces an excellent fuel wood. It is one of the most important tree species of Social Forestry Programmes in India. It has a very wide distribution right from the semi arid regions of Rajasthan to Tamil Nadu and in eastern most part of Bihar and Orissa.
- It also grows in riverain beds. Every part of the tree is used in one or the other way by the villagers. Example : *Acacia nilotica* in tank beds of Tamil Nadu.

4. Roadside Plantations

- The plantations established along the roads with the primary objective of providing much needed shade to the travelers, particularly during the summer months are called roadside plantations.

Objectives

The roadside plantations are required to meet the following objectives

i. Provision of comfort to the travellers

ii. Aesthetics and landscape improvement

iii. Stabilization of roadsides and checking of the shifting of sand to the road from adjoining areas in the desert

iv. Improvement of ecological conditions

v. Maximization of the productivity of the site to meet the requirement of the local people.

Design of Roadside Planting

- The planting systems usually followed are of the following types : viz., balanced line, unbalanced continuous line, unbalanced discontinuous line, sporadic system and parkway system. Of these, the first two are most widely practiced.

- The balance line system produced with continuous green wall of uniform sizetrees. Unbalanced continuous line produced as a result of alternating avenues of different species interspersed by ornamental trees should be more desirable.
- It should also be ensured that the avenues should not shut out of vie the spots of scenic beauty, important land marks and places of tourist interest, lakes and canals, beautiful hill or country features and snow peaks, etc. At some selected spots the parking system of planting may be adopted to develop picnic spots or resting places for the travelers.

Orientation

- The position of the first row of trees and the number of rows to be planted on either side of the road depend upon the category of the road. Single row planting is normally done along village roads and other district roads while more than one row can be planted in the case of highway roads.
- In the case of multiple row planting the first row is generally of shade and/ or ornamental trees and the remaining rows may be planted with fast growing trees on short rotation to produce tree products to meet the requirement of the local people.
- The spacing of the first and also of the subsequent rows may need adjustment with respect to the position of telephone or electric lines or drains or other structures along the road.
- Trees should not be planted under electric or telephone lines. Bushes and short ornamental trees can, however, be planted in such places. Trees in the first row are normally planted at least 6m away from the electric or telephone line.

Spacing

- The spacing may vary from high density to low density i.e. 1m × 1m to 5m × 5m depending on the space availability and based on the choice of species.
- The spacing between the trees in the first row of shade or ornamental trees should be wider than that between the trees in the remaining rows. The crown spread of the species to be planted determines the spacing.
- For shade trees normally having broad crowns a spacing of 12 to 14 m is suggested. While too wide spacing may defeat the object of providing shade to the travelers, too close spacing may adversely affect the visibility. Planting of shade trees in the first row at wider spacing may result in patchy avenues as a result of the mortality. It may,

therefore, be desirable to keep the initial planting closer in order to provide for mortality and selection of better trees for final retention.

Planting

- Planting should be done during monsoon season preferably from September to December. Planting of tall seedlings of 6 - 8 ft is considered ideal for planting along the roads. Deep planting by digging 60 cm^3 pits is essential for good survival and establishment. Watering should be mandatory till their establishment.
- Planting at about half the desired spacing is, therefore advocated and when the crowns of trees begin to touch, less promising ones should be thinned to leave the remaining trees at the desired spacing. Too close spacing along hill roads particularly on the curves is dangerous.
- The trees should not obstruct the view at places where the pedestrians or domestic animals might usually be crossing the road near the villages.

Choice of Species

- The choice of species is broadly determined by the climate, site conditions, object of planting, intensity of damaging agents, etc.
- The site conditions vary depending upon the soul and the modifications brought about in it in the course of the construction of the road.
- Removal of topsoil for road construction leaves only subsoil and also creates waterlogging conditions in many areas as the level of the roadside strip becomes lower than the adjoining land and road and at many places the construction of the road obstructs natural channels of drainage.
- Choice of the species has to be kept in view all such modifications brought about in the site. Further, the choice of species is determined by the objectives of avenue planting whether for shade, ornamental or production of tree products needed to meet the local demand.

Example : *Ailanthus excelsa, Albizia lebbeck, Albizia procera, Anthocephalus chinensis, Azadirachta indica, Dalbergia sissoo, Ficus glomerata, Kigelia pinnata, Madhuca longifolia, Mangifera indica, Melia azedarach, Mimusops elengi, Peltophorum pterocarpum, Pithecolobium saman, Pongamia pinnata, Samanea saman, Spathodea campanulata, Syzygium cumini, Tamarindus indica, Terminalia arjuna, Terminalia catappa.*

Benefits of social forestry

Social forestry as a name indicates it satisfies all the demands from the society-

Major economic benefits from the social forestry programmes

(i) **Fuel and energy requirements:** Social Forestry plantation can help meeting the energy requirement of local people. In India, almost 50-60% of population lives in rural areas which require huge volume of fuel for domestic purposes. Thus, social forestry could provide enough firewood and reduces pressure on natural forests.

(ii) **Fodder requirements:** Social forestry plantations also provide fodder which is beneficial to sheep and goat farming. The species like *Acacia, Prosopis, Terminalia*, Neem, *Albizia* introduced in social forestry are acting as excellent fodder for livestock.

(iii) **Cottage industries:** Social forestry plantations provide a wide range of raw materials which may be used in several cottage industries like basket making, furniture making, agriculture implements, rafters, poles, post, matchstick, pencil making, sports goods etc. which provide new term for rural income and employment activities.

(iv) **Forest based industries :** Social Forestry plantations act as an excellent source of raw material for several wood based industries like pulp and paper, resin, gum, honey, lac, oleoresins, edible oil, non–edible oil, katha, kutch, plywood etc. to increase socio-economic profile of the society.

(v) **Employment:** Social Forestry provides increased potential of employment particularly to the poor and landless labourers. Plantation and nursery activities provide bulk activities to unskilled and semi skilled labourers in the society. Cottage industries and forest based industries provide enormous opportunities and help to promote the livelihood security of rural populations.

(vi) **Regulation of water cycle:** Social Forestry Plantations helps to regulate the water particularly in the catchment areas. The trees act as a store house for many watersheds by its natural ability of trapping the river water.

(vii) **Flood control :** Social Forestry plantations help to reduce the surface runoff and also reduce the quantity of silt brought down by rivers and reduces the rise of river bed level. The force of flood water is broken by the trees which act as barriers and helps to reduce the floods.

Constraints in social forestry

Though, the science of social forestry is for the people, by the people and to the people, it also experiences the following constraints.

1. People are associated as long as the project is implemented. Once a project is closed, the people ignore the project which leads to failure of project.

2. Land ownership is another constraint experienced which resulted in failure of the project.
3. Lack of legal constitution is also identified as one of the constraints.
4. Long gestation of trees is one of the constraints.

SOCIAL FORESTRY PROGRAMMES

Social forestry programmes in India has taken its inception in the year 1950 through the festival of tree planting and got strengthened after the enunciation of National Forest Policy 1952. Several states in the country have implemented social forestry projects. Among them Gujarat is the first state in India to create a separate social forestry wing as early in 1970's. Other states like Tamil Nadu, west Bengal, Uttar Pradesh, Odissa, Punjab, Haryana, Rajasthan, Madhya Pradesh and Himachal Pradesh have also started intensive social forestry programmes with financial assistance from the following international agencies.

Sponsoring agencies for social forestry projects

1. World Bank
2. Swedish International Development Agency (SIDA)
3. United States Agency for International Development (USAID)
4. Canadian International Development Agency (CIDA)

The major social forestry projects and programmes along with the budget outlay and project activities are furnished in the Table 1.

Table 1. Major social forestry programmes in India

S.N.	Name of the project Project activity and state	Donor agency	Project period	Project cost in lakhs	Project activity	
					Farm forestry	Village woodlots
1.	Tamil Nadu Social Forestry Project	Phase-I SIDA Phase-II SIDA	1981-82 to 1985-86 extended to 1987-88 1988-89 to 1992-93	591.38 854.00	P 85,165 F 50.80 P 18,000 F 154.40	131,405 178.94 56,300 247.50
2.	West Bengal Social Forestry Project	World Bank	1981-82 to 1986-87 extended to 1989-90	348.65	P 52,000 F 18.84	6,000 17.84
3.	Maharashtra Social Forestry Project	USAID	1982-83 to 1989-90	564.00	P 44,035 F 257.74.	33,975
4.	Haryana Social Forestry Project	World Bank DAN I DA	1982-83 to 1986-87 extended to 1989-90	333.25	P 30,000 F 25.65	12,000 46.39

S.N.	Name of the project Project activity and state	Donor agency	Project period	Project cost in lakhs	Project activity	
					Farm forestry	Village woodlots
5.	Jammu & Kashmir Social Forestry Project	World Bank DAN I DA	1982-83 to 1986-87 extended to 1989-90	237.40	P 19,000 F 17.40	5,000 17.00
6.	AP Social Forestry Project	CIDA	1983-84 to 1987-88 extended to 1989-90	383.78	P 108,100 F 68.78	25,000 61.03
7.	Orissa Social Forestry Project	Phase I SIDA Phase-ll SIDA	1983-84 to 1987-88 1988- 89 to 1992-93	281.70 783.40	P 26,500 F 11.60 P 62,000 F 253.60	32,076 na 52,500 222.30
8.	Karnataka Social Forestry Project	World Bank ODA (U.K.)	1983-84 to 1987-88	552.30	P 120,500 F 180.00	26,946 98.50
9.	Kerala Social Forestry Project	World Bank	1984-85 to 1989-90	599.11	P 69,200 F 162.25	14,100 na
10.	National Social Forestry Project UP	World Bank USAID	1985-86 to 1989-90	1,611.60	P 147,210 F 536.20	14,000 150.40
11.	National Social Forestry Project Gujarat	World Bank USAID	1985-86 to 1989-90	1,296.50	P 230,500 F 143.60	35,000 369.80
12.	National Social Forestry Project HP	World Bank USAID	1985-86 to 1989-90	572.90	P 66,838 F 139.30	41,000 201.20
13.	National Social Forestry Project Rajasthan	World Bank USAID	1985-86 to 1989-90	391.90	P 91,500 F 121.80	5,000 30.20
14.	Bihar Social Forestry Project	SIDA	1985-86 to 1990-91	538.57	P 71,750 F 156.28	30,750 21.52
TOTAL				**F 9940.44**	**P 1,242,293** **F 2,090.50**	**508,730** **1662.62**

CHAPTER - 12

JOINT FOREST MANAGEMENT (JFM)

Since the enunciation of National Forest Policy 1988, forest management in the country has witnessed paradigm shift from production to conservation based management. This has also witnessed participatory mode of forest management and ensures participation of the village community in all decision making process. Besides enjoying the beneficial role of forestry, this JFM concept has been incorporated in several academic programmes and hence this chapter is conceived and presented with following sub themes.

- Definition
- History and evolution
- Objectives
- Programmes
- JFM success Stories
    - Arabari experience
    - Behroonguda experience

JFM - DEFINITION

Scholars and policy makers have defined Joint Forest Management (JFM) in different ways-

i. 'JFM is a concept of developing partnerships between fringe forest user groups and the Forest Department (FD) on the basis of mutual trust and jointly defined roles and responsibilities for forest protection and development'.

ii. 'JFM is a forest management strategy under which the Government (represented by the Forest Department) and the village community enter into an agreement to jointly protect and manage forestlands adjoining villages and to share responsibilities and benefits'.

Principle

- The fundamental principle of Joint Forest Management (JFM) is to link local people in the forest developmental activities by suitably involving local people in each and every stage of decision making process.
- It is an approach and programme initiated in the context of the National Forest Policy of 1988 wherein the state forest departments support local forest dwelling and forest fringe communities to protect and manage forests and share the costs and benefits from the forests with them.
- The key element in JFM is that communities have the power to manage the use of forests by members and also exclude non-members. The benefits to them is direct access and control on the use and sale of most NTFPs and a share in the income from timber as well as other intangible benefits from local ecosystem services – like water recharge, pollination, wildlife habitat etc.
- Thus involvement of communities in conservation of forests and wildlife is of paramount interest.

HISTORY, ORIGIN AND EVOLUTION OF JFM

- Though the joint forest management has a very long history of evolution but got attaraction only after 1970's. This due to the successful establishment of a community-based forest conservation movement initiated in Arabari in Midnapur District of West Bengal State during 1971-72.
- In Arabari, the movement was triggered by a Silviculturalist of the State Forest Department, who by offering incentives to local communities induced them to protect and regenerate degraded Sal forests.
- The local communities, were organized into 'Forest Protection Committees' (FPCs), successfully protected the degraded forests from illegal felling, overgrazing, fire, and encroachment.
- The second evidence was due to the grassroots movement in the Sukhomajri village in Haryana in the 1980s, to rejuvenate forests and agricultural systems in the village.
- In Sukhomajri, Haryana State, the movement started from amongst the people. Construction of earthen dams stabilized agricultural output in the village. Forests in and around Sukhomajri village regenerated as a result. The regenerated forests in turn provided valuable biomass, including bhabar grass to local communities.
- These instances attracted the policy makers at the Central Government to the need to go beyond the legalistic "Forest Conservation Act 1980".

- Institutional measures for arresting the alarming trend of forest depletion in the country were actively considered. The National Forest Policy (1988) enunciated was accordingly enacted by the Government of India with a strong focus on conservation, environmental stability and ecological balance through association of tribal's and local communities in protection, regeneration and development of forests.
- In pursuance of the National Forest Policy of 1988, the Government of India issued a 'circular' in June 1990 for involvement of village communities and village associations (VAs) in the regeneration of degraded forest lands. This marked the birth of the Joint Forest Management (JFM) movement in India.
- On 1st June 1990, Government of India, issued guidelines to the state governments for involving local communities in the protection and development of the degraded forests. Based on this, the JFM programme has now been widely adopted in twenty two states of the country.

Objectives of JFM

- Provide a visible role to the local communities in planning, management and protection of forests and to give them a share in the benefits from these forests.
- Elicit active participation of villagers in (a) creation (b) management and (c) protection of plantations.
- Achieve ecological needs consonant with sustainable productivity of wood and other non-timber forest resources.
- Wean away the land owning communities from shifting cultivation by adopting an alternative i.e. Tree farming.
- Productively utilize the degraded jhum-land thereby checking soil erosion.
- Conserve Biodiversity through people's action.
- Create and generate forest –based economy for the villagers.

Salient Features of JFM

- The JFM encourages development of partnership between local people and forest department to manage forest lands jointly.
- The JFM provides a common structure for consultation and negotiations between the forest department and the local people and assures legalized access for the local communities to forest lands.
- It facilitates benefit sharing mechanism and ensures livelihood opportunity for the local community.

- The JFM process encourages local people to protect forest from fire grazing and other illegal activities.
- The JFM process facilitates involvement of local people in all decision making process.

Structure of JFM

The structures and nomenclatures of JFM have varied from State to State. In general, one can delineate the following structure for JFM. The base of the JFM structure, which comprises village level institutions, is referred to as "forest protection committees". These bodies include a "General Body" and a "core", "Executive" or "Management Committee (MC)" elected by the General Body for discharging the assigned functions. The General Body comprises eligible members of the village. The eligibility criteria for membership vary from State to State.

In most states, adults are eligible for joining the General Body. The General Body elects the local community representatives in the Executive Committee. The Executive Committee has elected members ranging in numbers from 5 to 15. It also includes ex-officio, non-elected members drawn from the Forest Departments, local NGOs, village schools, village administrative and development officers and in some cases representatives of the Gram or Mandal Panchayats. The term of the Executive Committee is generally for two to five years in most of the States. The Executive Committee undertakes decision-making, planning and implementation of management plans.

Powers of Forest Protection Committees

The functioning of FPCs in the post-1990 phase indicate that there are widespread inter- State differences, when it comes to grant of execution of powers by Forest Protection Committees. Different States give different flexibilities to FPCs in the matter of administrative, executive and financial powers. Nevertheless, the standard parameters of power granted to forest protection committees can be broadly categorized as under-

- Framing of rules governing management of forests.
- Administrative powers for convening meetings of General Body and Management Committee meetings.
- Voting rights in Management / Executive Committees.
- Punitive powers – i.e., powers for punishing and imposing fines on offenders.
- Rights to collect specific resources and sell them.
- Distribution of benefits arising from conservation and regeneration.
- Cancellation of membership of recalcitrant members.
- Financial powers – to incur expenses, maintain accounts.

JFM Methods

(a) Participatory Rural Appraisal for Joint Forest Management-tools and techniques

Participatory Rural Appraisal (PRA) tools facilitate collection and analysis of information by and for community members. PRA emphasizes local knowledge and involves communities in the inventorying, monitoring, and planning of local forest management. Because it is a collaborative process, PRA actively empowers marginalized communities, de-emphasizes hierarchies, and helps to identify resource needs and sustainable use systems. Each of the tools provided in this manual is based on PRA. PRA methods serve multiple purposes. They provide information to outsiders who wish to understand how the community uses and manages its resources and they provide information for the collective community to evaluate its resource management practices. The process of collecting PRA information is as important as the data itself, as it triggers dialogues with the community, foresters, NGOs, and local government to examine existing resource use practices, problems, conflicts, and opportunities, providing a basis for developing more sustainable and productive management systems.

(b) Dialogue process

(i) Semi-structured interviews

This technique involves obtaining valuable information through simply talking to individual community members or small groups. It is not limited to gaining stakeholder information but is used to better understanding the community's relationship and dependency on the forest resources, the management issues, and the available resources in addition to a whole range of additional information.

(ii) Group discussion

Group discussions with the community or multiple stakeholders allow us to explore a range of concerns and interests in the same way the semi-structural interviews do. The process of group discussions raises awareness of resource concerns and conflict and provides a platform to negotiate issues. The information that is gathered during the group meeting might help develop a management plan that is acceptable to all forest user groups.

(iii) Diagrams & Lists

Venn diagram

Venn Diagrams consist of circles drawn to indicate different user groups (overlap to reflect common membership). Identify leaders or contact representatives within the larger circles. Venn diagrams are useful in illustrating forest user groups and the relationship and reliance each group has with the forest and each other. They can help identify conflicts over resources as well as establish which village institutions might play a lead role in supporting a participatory management.

System diagram

This technique is a way to illustrate the user groups (including village communities, government field staff, commercial interests, NGO researchers) and the interaction between user groups and the resource.

Seasonal Calendar

The seasonal calendar documents the flow of forest products over time and how product collection changes over the season. This information can contribute to an estimation of yields.

Flow diagram

This diagram technique is one of a range of tools that can be used to develop an economic assessment of the production system. This along with other economic techniques to estimate yields, labor and capital costs draw on information obtained from PRA tools but require additional research and ecological data before a rigorous economic analysis can be achieved.

Historical transect

The technique of having the community construct a transect of the area helps to identify successful and unsuccessful management systems so that a new management system can avoid the same mistakes and promote the positive aspects endorsed by the community. The decline and or increase of certain forest products and species can be established such that the management plan can address these issues.

Trend lines

Creating a graph that diagrams changes in the population, rainfall, and volume flow of important forest products over time is important. Trend lines can also be used to chart patterns of forest disturbance and regeneration over time.

(iv) Tables & charts

List making

This collaborative process allows us to (a) identify the forest products and (b) classify forest products into commercial and subsistence use. Classifications represent use categories or use types. Use types may include: edible flora (mushrooms, fruit, seeds, nuts, etc); edible faunas (insects, honey, fish, animals, etc); construction material (timber for walls, grasses for ropes, etc); medicinal (leaves, bark, etc); fuels (trees, shrubs, leaves, etc); fodder, (trees, grasses, etc); and others (gums, resins, lac etc.)

Organizational chart

These illustrate the structures and individuals within local government organizations and technical agencies. Organizational charts can also be used to reflect indigenous organizations including village chiefs, council of elders, healers, water management organizations and leadership positions, etc. However, because traditional community leadership and institutional patterns are not linear, other Visual tools like Venn diagrams may be more effective in reflecting their formations.

Time line

Time lines help identify important past events. For example, drought, forest felling, forest fires, etc. This information assists the PRA team in being better informed about the area and the potential risks posed to the natural resources. Understanding how the community dealt with past events may help the team to facilitate discussion and select suitable PRA methods.

Ranking

This process indicates the order of importance among the forest products that are identified by the community. For less complex issues, villagers can be asked to rank products during semi-structured interviews. For more complex issues, it is helpful to develop a pair-wise ranking system or scoring (matrix ranking) system. It is useful to compare forest products within a given type of use, i.e. fuel wood, construction timber, etc. It is also useful to limit the number of forest products/ species to be ranked or scored. In order to limit the species/products to be ranked / scored, villagers should be asked to identify the most popular species within a use type. All commercial species/products should also be included.

Scoring (Also Known As Matrix Ranking)

This process helps identify the relative importance of each forest product for the household economy. To aid comparison, it is useful to allow villagers to compare forest products within each type of use, i.e. fodder, fuel, food, etc.

(v) Spatial analysis

Community sketch mapping

Participatory mapping techniques involve facilitating community members in developing spatial representations of their areas by creating maps on the ground or on a large piece of paper. Such maps reflect the locations of villages, forests, agricultural land, water resources, as well as management issues. The process of making the map and the discussions that occur while it is being made are important outputs of the exercise. Map information can be transferred to a paper and digitized so that it is documented for future reference.

Sketch maps provide a rapid visual representation of the resource system that is easily understood by villagers and foresters. Sketch maps also provide a means to identify the sources of forest products but also can be used to gain better information on the forest's conditions and community use patterns. This information can then be used to supplement the spatial analysis of the management area, i.e. where is the management area under discussion?

Benefits of JFM

(a) The JFM has institutionalized the people's involvement in each and every stage of decision making process and help to conserve the degraded forest.

(b) It Empowers the people which resulted in active participation of communities in restoration, conservation and management of local forest.

(c) The JFM helps to create awareness of the need to manage environmental resources like soil, water, air, etc.

(d) The JFM activities rehablites the degrade forest and enriches the biodiversity of the localities. it also improves the quality and quantity of forest produces and accrues the benefits.

(e) JFM develops a new method of learning experience for governance. The decentralization of implementation and involvement of local community have not only reduced the burden of the forest department but also help to provide a learning experience throughout the process of implementation.

(f) The JFM created Adaptive approach which resulted in establishment of women self help group, equitable distribution of benefits, reducing the poverty line, increasing the water table, improvement in health sanitation etc.

(g) The JFM activities helped the communities and forest department to develop better understanding of the requirements of natural resource management.

Important Features of Policy Directives on Joint Forest Management (JFM)

Policy Directives	Main Features
The Circular (first) Concerning JFM, 1990	• Involvement of village communities and NGOs in regeneration of degraded forest. • Benefits of people's participation should go to the village communities.

Policy Directives	Main Features
JFM Cell Creation Notification, 1998	• Cell created for monitoring impact of JFM being carried out by state government.
Standing Committee Notification, 1988	• Advise on all operational aspects of JFM for expansion of JFM to non-forest area.
Terms of Reference Notification, 1999	• Sharing of experiences of JFM implementation as each state has passed its own resolution on JF M • Monitoring of JFM programme.
Notification for JFM Network, 2000	• Creation of `JFM Network' at the national level to act as regular mechanism for consultation between various agencies involved in JFM • Representatives from the entire stakeholder.
Guidelines for Strengthening JFM, 2000, 2002	• JFM policy directives and broad framework for implementation of JFM. • Measures such as legal support for JFM Committees, promotion of women's participation and conflict resolution. • Memorandum of Understanding between forest department and JFM Committees outlining short- and long-term roles and responsibilities, and pattern of sharing of usufructs. • Capacity building for management of non-timber forest products for providing remunerative prices for user.
Operation Guidelines for Tenth Five Year Plan, 2002–2007	• Formulation of National Afforestation programme to encourage participate approach for sustainable forest development. • Implementation of centrally sponsored afforestation schemes via a two-tier system consisting of Forest Development Agencies and JFM Committees. • Transfer of funds to JFM Committees through Forest development Agencies.

Strengthening of Joint Forest Management

The JFM programme in the country was reviewed by Government of India from time to time in consultation with State Governments, NGO's and other stake-holders in view of several emerging issues. In order to further strengthen the programme, the State Governments may take action on the following suggested lines.

(A) Legal backup to the JFM committees

i. At present, the JFM committees are being registered under different names in various States as per the provisions contained in the resolutions. Except in a few States where the committees are registered under the relevant acts in most of the states there is no legal back up for these committees. It is therefore, necessary that all the State Governments register the JFM or village committees under the Societies Registration Act, 1860 to provide them with legal back up. This may be completed by 31st March, 2000. Completion of such formation of existing JFM committees may please be reported to this Ministry.

ii. There are different nomenclatures for the JFM committees in different States. It would be better if these committees are known uniformly as JFM committees (JFMC) in all the states. Memorandum of Understanding, with clearly defined roles and responsibilities for different work or areas should be separately assigned and signed between the State Governments and the committees. All adults of the village should be eligible to become members of the JFM Committees.

(B) Participation of women in the JFM programme

Considering the immense potential and genuine need for women's participation in JFM programme, following guidelines are suggested for ensuring meaningful participation of women in JFM.

(i) At least 50% members of the JFM general body should be women. For the general body meeting, the presence of at least 50% women members should be a pre requisite for holding the general body meeting.

(ii) Atleast 33% of the membership in the JFM Executive Committee/ Management Committee should be filled from amongst the women members. The quorum for holding meeting of such Executive/ Management Committee should be one-third of women executive members or a minimum of one whichever is more. One of the posts of office bearer i.e. President/ Vice-President/ Secretary should be filled by women members of the Committee.

(C) Extension of JFM in good forest areas

- For better resource planning and collective management distance from the village and dependency on forests should be the main criteria for allowing JFM programme to operate.

- Therefore, JFM programme should cover both the degraded as well as good forests (except the protected area network). The micro plan or treatment plan and memorandum of understanding should be different for degraded forests and good forests (crown density above 40%).
- In good forest areas, the JFM activities would concentrate on NTFP management and no alternation should be permitted in the basic silvicultural prescription prescribed in the Working Plan but to promote regeneration, development and sustainable harvesting of NTFP which can be given free or on concessional rates as per existing practice in degraded areas under JFM.
- The benefit sharing mechanism will also be different for the good forest areas. The JFM committees will be eligible for benefit sharing for timber, only if they have satisfactorily protected the good forests for a minimum period of at least 10 years and the sharing percentage should be kept limited to a maximum of 20% of the revenue from the final harvest.
- The felling of trees and harvesting of timber will be as per the provisions of the working plan. A certain percentage of revenue from final harvest should be ploughed back in the silviculture & management of the forests.
- The extent of good forest areas to be allowed will depend upon the number of village household and should be restricted to a maximum limit of 100 ha and generally limited to 2 km from the village boundary.
- For degraded forests also as far as possible JFM should be first concentrated on areas up to 5 km from the village boundary.
- The implementation of JFM in good forest areas shall be done in a phased manner on pilot basis.
- The pilot areas may be monitored closely for a few years and based on the feedback and success achieved the programme can be extended further in consultation with the Central Government.
- Before allowing the good forests on pilot basis, all the degraded forests of that locality should be covered simultaneously.

(D) Preparation of micro plan in JFM areas

i. In case of new working plans a JFM overlapping working circle should be provided to incorporate broad provisions for micro plans. To achieve these flexible guidelines should be evolved for preparation of local need based micro plans. For this purpose, the working plan officer will work in tandem with the territorial DFO and CF for finalization of the prescriptions of the JFM overlapping working circle. The micro plans should be prepared by the Forest Officers and Village Forest Protection Committees after detailed PRA exercise and should reflect the consumption and livelihood needs of the local communities as well as provisions for meeting the same sustainably. It should utilize locally available knowledge as well as aim to strengthen the local institutions. It should also take into account marketing linkages for better return of NTFPs to the gatherers and should also reflect the needs of local industries/markets. This should be done with due regards to the environmental functions and productive potentials of the forests and their carrying capacity as also their conservation and biodiversity values.

ii. In areas where the existing working plans are in force (till their revision in future), for incorporation of micro plans in the working plans, a special order may be issued by the PCCFs for implementation of the micro plan. In these areas, micro plan should aim at ensuring a multi product and more NTFP oriented approach. Without changing the basic principles of silviculture, deviations may be approved in the existing working plans if necessary. To ensure this, the concerned DFO and CF should dovetail the requirements of micro plans with the working plans.

iii. The micro plan should also take into consideration and provide suitable advice for areas planted/ to be planted on community lands and other Government lands outside the notified forest areas including in the district council areas of North East.

iv. Infrastructure/Eco-development under micro plan should form a separate entity for funding it through concerned development agencies.

(E) Conflict resolution

In order to resolve conflicts in the functioning of JFM committees and to maintain harmony among different groups participating in the JFM, State Governments may constitute divisional and state level representative forums or working groups. This forum/ group should include representatives from all the stakeholders including NGOs. The model prescribed by the Andhra Pradesh Government for this purpose is a case in point for consideration.

(F) Recognition of Self-initiated groups

The community groups in many places in Orissa, Bihar, Gujarat, Andhra Pradesh and Karnataka are performing the essential functions of forest protection and regeneration. These groups need to be identified, recognized and registered as JFM Committees after proper verification of records and enquiry. The period of their existence and duties performed for protection and regeneration should be suitable assessed and proper weightage given to them for deriving benefits under the JFM programme.

(G) Contribution for regeneration of resources

For long term sustainability of resources, it is essential that a mechanism is created for ploughing back a certain percentage of the revenue earned from final harvest. For this purpose, no less than 25% of the share of village community should be deposited in the village development fund for meeting the conservation and development from its share of such sales. There should be transparent mechanisms for computation of income for sharing the benefits between different stakeholders.

(H) Monitoring and evaluation

Concurrent monitoring of progress and performance of this programme should be undertaken at Division and State Level. Evaluation of the programme should be planned at an interval of 3 years and 5 years at Division and State level respectively.

SUCCESS STORIES OF JFM IN INDIA

a. Arabari experience

The Arabari forest range of Midnapur Forest Division, West Bengal has experienced complete degradation of natural forests particularly the sal forests and the associated natural regeneration. The local people have exploited the area which leads to loss of genetic resources. Hence the participatory approach of forest management has started its inception in Arabari by the then DFO Shri A.K. Baneerji. He has earmarked 1272 ha area of degraded forests for this approach and successfully established this area with plantations through the participation of local dwellers. Under this participatory approach the villages were provided with employment generation activities coupled with exclusive rights to collect all non wood forest products at free of cost. The villages were allowed to cultivate paddy, fodder, grasses, maize and groundnut. They have inturn protected the forests by establishing forest protection committees. This has shown excellent protection and the forests have been regenerated which formed the basis for implementation of JFM in the country.

b. Behroonguda experience

Behroonguda is one of 77,000 hamlets and villages in Andhra Pradesh. The 97 families who belong to the Gond and Naikpod tribes formed themselves into a forest protection group in 1990, but the state government officially recognized their efforts in 1993. In 1998, five years after the recognition of the committee - referred to as Vana Samarakshana Samithi, or VSS - Behroonguda residents began to derive usufruct benefits from the forest. The costs they have borne and the benefits they have derived from protecting 500 hectares of degraded forest allotted to them as part of the joint forest management (JFM) in Andhra Pradesh. The VSS is widely regarded as being successful.

An analytical method is developed and applied to two case studies of communities managing a mixed teak forest system and a sal coppice forest system in Madhya Pradesh the analytical results show that-

a. Economic returns to JFM are considerable in both forest systems;

b. There is an increase in revenues from the forest to the communities but a theoretical loss to forest departments;

c. Income flows into the communities increase significantly, partially offset in the sal coppice system by declining income from collection of non-timber forest products (NTFPs);

d. JFM benefits are not always distributed equally, resulting in collectors of firewood and some NTFPs losing, though overall gains are sufficient to compensate losers.

■■■■

CHAPTER - 13

DENDRO ENERGY RESOURCES

The growth and development of any nation is valued based on its energy production and consumption. Though India is continued to be energy starving country but satisfies the demand both through its own generation and by imports. The per capita consumption of power in the country is growing at an alarming rate. Currently over 50% of the energy generation is met from fossil fuel like coal and is one the causative factors for the growing environmental problems. Hence the government is mandated to generate energy from renewable resources which are green and clean. Among the renewable resources, the potential of the tree based dendroenergy has gained significant attraction as it is classified as a carbon neutral process. Hence the chapter on dendroenergy resources have been incorporated and presented with the following sub themes.

- Definition
- Use the Energy in Biomass
- Sources of Dendro biomass (Raw materials)
- Types of biomass power generation
- Thermo chemical properties of important energy species
- Benefits of dendro energy

BIOMASS ENERGY

When biomass is produced by plants as a result of photosynthesis, some of the energy in the sunlight is converted into the chemical energy that binds various atoms to form carbohydrate molecules. Herbivores (animals that feed on plants) eat the plants and make their own biomass by using the energy stored in the plants' biomass. Similarly, carnivores (animals that feed on other animals) consume the biomass of other animals and in the process make their own biomass. When the plants or animals die, the energy captured in their biomass remains until the biomass decomposes through natural mechanisms or is burned or processed by humans. Biomass energy is the utilization of energy stored in organic matter. It is humanity's oldest external source of energy, dating back to prehistoric man's first use of fire. And biomass is still an important part of the world's energy system; the use of traditional biomass charcoal, firewood, and animal dung in developing countries accounts for almost 10% of the world's primary energy supply.

- Biomass can be used to produce heat and electricity or used in combined heat and power (CHP) plants.
- Biomass can also be used in combination with fossil fuels (co-firing) to improve efficiency and reduce the buildup of combustion residues.
- Biomass has potential to replace petroleum as a source for transportation fuels.
- Biomass is also used in conjunction with fossil fuels for electricity generation in "waste-to-energy" projects. These are niche applications, which depend on the biomass having no other commercial value and being in close proximity to the application.

RENEWABLE BIOMASS

The capture of solar energy as fixed carbon via photosynthesis is the key initial step in utilising biomass:

$$CO_2 + H_20 + light + chlorophyll => CH_2O + O_2$$

This extremely important photochemical process as the plants can actually capture and store this solar energy. This stored energy can be then be used as a fuel source. Burning the biomass returns the CO_2 that was absorbed as the plants grew to the atmosphere and releases the energy. Biomass is commonly stated as having no net release of CO_2 if the cycle of growth and harvest is sustained (called carbon neutral). This is not strictly true as it takes some energy (and therefore a discharge of CO_2) to grow, harvest and/or transport the biomass for use as a fuel source. However the use of biomass fuels can result in the displacement of emissions of carbon dioxide ordinarily released using fossil fuels. This displacement will depend entirely on the efficiency with which the biomass energy can be produced and used.

The absorption and release rate of carbon is an important part of the Carbon Cycle, which is explained below:

Carbon Cycle

All living organisms (plants and animals) contain the carbon atom. Under normal conditions, carbon compounds can be solid, liquid, or gas. Carbon atoms continually move through living organisms, the oceans, the atmosphere, and the crust of the planet. Plants absorb CO_2 from the atmosphere during photosynthesis (called primary production) and release CO_2 back into the atmosphere during respiration. Another major exchange of CO_2 occurs between the oceans and the atmosphere. The dissolved CO_2 in the oceans is used by marine biota in photosynthesis. This movement is known as the carbon cycle. The paths taken by carbon atoms through this cycle are extremely complex, and may take millions of years to come full circle.

USE THE ENERGY IN BIOMASS

There are a wide variety of uses for the energy in biomass, including basic life functions, direct combustion, and charcoal, liquid fuel, or gaseous fuel production.

(a) Use for Basic Life Functions

We are all familiar with the use of biomass for basic life functions by all animals and humans. All the food we eat contains biomass, whether vegetables, animals, or products derived from them. The energy in these foods is what sustains life in all living organisms. Considerable debate exists about the competing uses of biomass for food versus energy production.

(b) Direct Combustion of Biomass

Most biomass is in solid form and can be burned (combusted with oxygen in air). The combustion process results in the generation of heat (thermal energy) the same way that burning coal gives off heat. The heat generated by the burning of biomass can be used for space heating (e.g., heating of buildings), for cooking, and for heating water. It can also be used to boil water to generate steam, which in turn is used to run a turbine and electric generator to produce electricity in the same way that coal is used to generate electricity in a coal-fired power plant. The biomass may need a certain amount of processing before burning (e.g., sorting, drying, and size reduction). It may also be mixed with coal or other types of fuel in the furnace.

(c) Charcoal Production from Biomass and Combustion

In addition to water (moisture), wood normally contains chemicals that volatilize when they are heated. During the combustion of wood in air, most of this volatile matter oxidizes and contributes to the energy being released in the combustion process. However, when wood is heated in the absence of air or oxygen, the volatile matter is driven off of wood in a process called pyrolysis, and the remaining matter is called charcoal. Charcoal is almost pure carbon, with about twice the energy content per unit mass as the original wood. Therefore, it can burn at much higher temperatures than wood. However, it takes about 4 to 10 kg of wood to make 1 kg of charcoal. In addition, if the volatile gases driven off from the wood during pyrolysis are not collected and used, they contribute to greenhouse gas emissions.

(d) Production of Liquid Fuels from Biomass

Liquid fuels that are produced from biomass are used to extend crude oil resources by substituting for gasoline or diesel fuel. Although there are different ways of producing gasoline-like liquid fuels from biomass, the most common gasoline substitute is ethanol. The common diesel substitute is biodiesel. The production of ethanol and biodiesel from biomass are discussed below.

Ethanol Production from Biomass

The most common method for producing ethanol from biomass is through a process known as fermentation, which is an anaerobic biological process in which sugars are converted to alcohol by the action of microorganisms such as yeast. Fermentation is a technology that has been used for centuries to make beer and wine. The process requires sugars. If the carbohydrates contained in the starting biomass are starch, such as is found in corn, then the starchy material first has to be converted to sugars. The most common crop used in the United States for making ethanol for gasoline additive is corn, whereas in tropical countries such as Brazil, the crop of choice is sugar cane. The amount of ethanol produced per unit mass of biomass, given in units of liters per kg (L/kg), varies from about 0.07 for sugar cane, 0.16 for wood, and 0.36 for corn. The annual yield per unit land area varies substantially from place to place, but generally it ranges from about 400 to 12,000 L/ha/yr for sugar cane, 160 to 4,000 L/ha/yr for wood, and 250 to 2,000 L/ha/yr for corn.

Biodiesel Production from Biomass

Biodiesel is normally produced from vegetable oils, animal fats, or greases. The process involves the use of a catalyst and alcohol. The catalyst is typically sodium hydroxide or potassium hydroxide, which is dissolved in methyl alcohol. Any type of vegetable oil or used cooking oil can be used to make biodiesel; most biodiesel today is made from soybean oil. Biodiesel production from algae is still in the experimental stage. Algae are expected to produce from 10 to 100 times more biodiesel compared with soybeans and other seed crops. Biodiesel is most often blended with petroleum diesel in ratios of 2% to 20%. It can also be used as pure biodiesel. Biodiesel fuels can be used in regular diesel vehicles without modifying the engines. They can also be stored and transported by using existing diesel tanks and equipment.

(e) Production of Gaseous Fuels from Biomass

It is well known that anaerobic digestion of organic matter produces organic gas, mostly methane. Examples of natural processes are the production of landfill gas in municipal landfills and the burping of cows. Biomass can also be converted to biogas under controlled conditions in what is known as a digester. The organic matter used for this purpose is often animal waste such as manure. The anaerobic digestion is basically a biological process carried out by bacteria in the absence of oxygen. Biogas can also be produced from solid biomass by chemical means similar to the way town (or coal) gas is produced from coal. Biogas can be used in many applications similar to natural gas, including cooking, space heating, industrial heat, and electricity production.

SOURCES OF DENDRO BIOMASS (RAW MATERIALS)

Raw materials that can be used to produce biomass fuels are widely available across the country and come from a large number of different sources, and in a wide variety of forms. All of these forms can be used for fuel production purposes, however not all energy conversion technologies are suitable for all forms of biomass. The biomass sources from forestry sector alone are furnished.

a. Wood and wood residues

Wood consists of wood and other products such as bark and sawdust which have had no chemical treatments or finishes applied. Wood may be obtained from a number of sources which may influence its physical and chemical characteristics. The lignocellulosic wood is suitable for a range of energy applications. It can be burned for heat and/or power at a range of scales. New 'second generation' technologies are being developed which are capable of producing a range of liquid or gaseous transport bio fuels from lignocellulosic wood.

i. Bark

Bark may be removed from saw logs and available as a residue from wood processing. Bark typically contains high levels of minerals and consequently is prone to give high levels of ash and slogging in combustion systems. It may, however, be a suitable fuel for generating process heat close to where it is produced, such as for firing drying kilns at a sawmill. Minerals will be retained in the ash and consequently this may be used as a soil fertilizer.

ii. Logs

Small Round Wood (SRW) may simply be cut into logs. This may be done in the plantation for ease of extraction and handling and to assist drying, and involve delimbing and cutting into logs of typically 2-3 m in length.

iii. Sawdust

Sawdust is typically available as a co-product of wood processing or manufacturing. It may be of high moisture content, e.g. from cutting green wood in sawmill, or very dry from furniture manufacturing. It may have a bulk density only 30% of that of the solid wood and so, even if very dry, has a very low energy density. It does, however, present an extremely large surface area to volume ratio and is suitable for blowing into some combustion or gasification systems. Sawdust, especially dry sawdust, is particularly suitable for processing into pellets.

iv. Wood chips

Although logs can be stored and transported conveniently when stacked, and the ease of air passage through a log pile allows good drying, they may not always be the most convenient form for automated handling and feeding. Also, the relatively small surface area to volume ratio is not ideal for efficient combustion or gasification. Wood chips can form a much more uniform fuel that can flow and can be fed to a boiler, gasifier or other conversion system as a steady flow using an auger feed or a conveyor. With a large surface area to volume ratio they can also be burned very efficiently. Wood chips may have a bulk energy density of about 50% of that of the solid wood.

v. Wood pellets and briquettes

Wood pellets are made from dry sawdust compressed under high pressure and extruded through a die. They may include a low level of added binder, such as starch, but many use nothing other than steam. Wood pellets should be dry, clean, mechanically robust and have an ash content defined by the appropriate standard to which they have been made, which may also define other contaminants such as chlorine content. Briquettes are similar to wood pellets, but physically larger. Sizes vary but briquettes can vary in diameter from around 50 mm to 100 mm+. Briquettes are usually between 60 mm and 150 mm in length. They can offer a cleaner, more consistent alternative to firewood logs, offering higher energy density and steady combustion.

vi. Wood from plantations

This includes wood from private and state owned forest and plantations. As harvested, wood will be at a range of moisture content, and of a variety of physical shapes and sizes. In addition to harvesting, some level of pre-processing is likely to be required. Transport and storage will also be necessary.

Wood processing industry co-products

The wood processing industries, such as sawmills and timber merchants are also a source of virgin wood in the form of off cuts, bark and sawdust. Many transport, pre-processing and storage issues may be similar to those for plantation products, however the material obtained is likely to be in different forms from plantation or arboricultural products.

There are likely to be a number of different output streams with different characteristics:

- Sawmill off cuts may include a high proportion of bark.
- Some material may have been kiln dried to extremely low moisture content making it potentially very suitable for wood pellet manufacturing or blending with material at higher moisture content.
- There may be sawdust at a range of moisture content from different stages of processing, and again very dry sawdust from kiln dried timber may be very suitable for wood pellet manufacture.

TYPES OF BIOMASS POWER GENERATION

The three primary routes for biomass to power are: Combustion, Gasification and Anaerobic Digestion. Combustion is easy to understand, instead of coal or other fossil fuels; biomass is used to produce steam that runs a turbine. Combustion of biomass for power could either be in the form of co-firing (when it is burned along with coal) or pure play biomass based combustion. In the case of gasification, the biomass is first gasified and this gas in turn produces power in a gas engine.

Anaerobic digestion is usually applied to biomass that typically has a high amount of water in them (anaerobic digestion is commonly used for treating organic waste such as kitchen waste and sewage waste into energy). Under this route, microorganisms act upon the organic matter present in the biomass under anaerobic (absence of air) conditions and convert it into biogas. An emerging route for biomass based power is pyrolysis. In this, the biomass is rapidly heated to 450 - 600°C in the absence of air, and results in a bio-oil called the pyrolysis oil, which in turn could be used for firing the boilers. Typically, 50 - 75% (by weight) of the feedstock is converted into pyrolysis oil.

a. Combustion

- Typically works well beyond 5 MW
- Well established technology works on the regular rankine cycle
- Comprises over 85% of installed capacity for biomass based power production in India (excluding biomass cogeneration)
- Works well for most types of biomass

i. Co-firing

This is a sub-set of combustion based power production. Some of the modern coalfired power plants use biomass for co-firing along with coal. It is quite efficient, cost-effective and requires moderate additional investment. In general, combustion efficiency of biomass can be 10 percentage points lower than for coal at the same installation, but co-firing efficiency in large-scale coal plants (35%-45%) is higher than the efficiency of biomass-dedicated plants. In the case of co-combustion of up to 5%-10% of biomass (in energy terms) only minor changes in the handling equipment are needed and the boiler is not noticeably de-rated. For biomass exceeding 10% or if biomass and coal are burned separately, then changes in mills, burners and dryers are needed. Many co-firing technology options have been demonstrated in several countries (Northern Europe, United States and Australia) in some 150 installations using different feedstock (wood biomass, residues and crops).

Effects of Co-firing on Plant Operation and Emissions

Co-firing can have several effects on the emissions and the plant operation: positive effects are that SO*X* and NO*X* emissions usually decrease due to the lower sulfur and nitrogen content in biomass than in coal. Furthermore, alkali components in biomass ash can have an effect of SO*X* removal. Since biomass has a high volatile content, it can also be used as reburn fuel for NO*X* reduction from the coal combustion, which gives a further potential for significant decrease of the NO*X* emissions. Negative effects of co-firing are additional investment cost for biomass pretreatment and boiler retrofitting, higher operation cost due to increased fouling and corrosion, and a possible decrease of the electric efficiency (if the superheater temperature has to be decreased due to high temperature corrosion). Besides potential poisoning of SCR catalyst also the efficiency of electrostatic precipitators may be reduced. Furthermore, the utilization of the ash and the residues from the flue gas cleaning system (especially the De-SO*X* installation) has to be considered when co-firing biomass. The ash quality can be negatively influenced mainly by alkali metals and chlorine contained in biomass. Furthermore, also the content of unburnt carbon can increase.

ii. Cogeneration

When bagasse is also used as a feedstock for steam generation in boilers in sugar mills, the main objective is process heating, and thus power production is only an ancillary benefit (which is why the term cogeneration – generation of both heat and power). As a result, the capacity of the power production is set based on the heating requirements. Thus, during the cane season, these mills have enough bagasse to be used for both heating and power production. During the non-cane season, typically coal is used for power production, as there are no heating requirements.

b. Gasification

In the case of combustion, the biomass is fired to generate steam which turns a turbine for power production. In the case of gasification, the biomass is gasified into a mixture of CO and H_2 in a gasifier. This gas is fed to a gas engine that produces electricity. Biomass gasification refers to the incomplete combustion of biomass resulting in production of combustible gases consisting of Carbon monoxide (CO), Hydrogen (H_2) and traces of Methane (CH_4). This mixture is called producer gas. Producer gas can be used to run internal combustion engines (both compression and spark ignition) for power production, or can be used as substitute for furnace oil in direct heat applications. Gasifiers can work at low scales – as low as 20 kW, and works well up to 2 MW, with current technology. Technology uses a combination of gasifier and gas engines. The technology has been in vogue for decades, but is still evolving. Currently, less than 125 MW of cumulative installed capacity in India (less than 15% of total biomass power, excluding biomass cogeneration). Works best for woody biomass, but latest gasifiers also work reasonably well with nonwoody (including fine biomass).

Biomass gasifiers are reactors that heat biomass in a low-oxygen environment to produce a fuel gas that contains from one fifth to one half (depending on the process conditions) the heat content of natural gas. The gas produced from a gasifier can drive highly efficient devices such as gensets, turbines and fuel cells to generate electricity.

Gas engines

Gas engines used in the gasification power plants run on producer gas to produce power. These engines have reasonable efficiencies. Producer gas has a very low air to fuel ratio (1.3:1) compared to other gases (Natural gas 17:1). Hence the engine has to be operated with a different carburetor for producer gas and natural gas. Also, the producer gas can be operated in an engine which has a higher compression ratio. Engines used for producer gas have generally lower in efficiency and they are cheaper. The efficiency of these engines is more than 30%.

Power evacuation system

Suitable evacuation systems need to be provided in order to transmit the electricity to the grid (if it is grid connected) or to the location where it is required (if it is captive).

Advantages of Biomass Energy

Biomass power has many advantages compared with other forms of renewable energy and is an important component of our nation's energy strategy. In many ways, biomass is a new source of power. While wood has always served as a fuel source for fires and ovens and conventional heating methods, biomass energy advancements are a few steps beyond that. Now these biomass fuel products are harvested and mass-produced and used in everything from engines to power plants.

Comparison of Biomass energy with other energy sources

Particulars	Biomass	Solar	Coal/Gas	Nuclear	Wind
Dispatchable/Baseload	√		√	√	
Availability	90%	15%	90%	90%	20%
Proven Technology	√	√	√	√	√
Cost-Effective	√		√	?	
Green	√	√			√

Clean Energy

Because of its relatively clean use, biomass energy, when used in commercial businesses such as airlines, receives tax credit in some countries like US. This is good for the environment and good for business. It does release carbon dioxide but captures carbon dioxide for its own growth. Carbon dioxide released by fossil fuels are released into the atmosphere and are harmful to the environment.

Abundant and Renewable

Biomass products are abundant and renewable. Since they come from living sources, and life is cyclical, these products potentially never run out, so long as there is something living on earth and there is someone there to turn that living things components and waste products into energy. In the United Kingdom, biomass fuels are made from recycled chicken droppings. In the United States and Russia, there are plentiful forests for lumber to be used in the production of biomass energy.

Reduce Dependency on Fossil Fuels

It has developed as an alternate source of fuel for many homeowners and has helped them to reduce their dependency on fossil fuels.

Reduce Landfills

Another benefit of this energy is that it can take waste that is harmful to the environment and turn it into something useful. For instance, garbage as landfill can, at least partially, be burned to create useable biomass energy.

Can be Used to Create Different Products

Biomass energy is also versatile, as different forms of organic matter can be used to create different products. Ethanol and similar fuels can be made from corn and other crops. With so many living things on the planet, there is no limit to how many ways it can be found and used.

Disadvantages of Biomass Energy

Besides above advantages, there are also some downsides to it. Below are some of its disadvantages.

Expensive

Firstly, its expensive. Living things are expensive to care for, feed, and house, and all of that has to be considered when trying to use waste products from animals for fuel.

Inefficient as Compared to Fossil Fuels

Secondly, and connected to the first, is the relative inefficiency of biomass energy. Ethanol, as a biodiesel is terribly inefficient when compared to gasoline, and it often has to be mixed with some gasoline to make it work properly anyway. On top of that, ethanol is harmful to combustion engines over long term use.

Harmful to Environment

Thirdly, using animal and human waste to power engines may save on carbon dioxide emissions, but it increases methane gases, which are also harmful to the Earth's ozone layer. So really, we are no better off environmentally for using one or the other. And speaking of using waste products, there is the smell to consider. While it is not physically harmful, it is definitely unpleasant, and it can attract unwanted pests (rats, flies) and spread bacteria and infection.

Consume More Fuel

Finally, using trees and tree products to power machines is inefficient as well. Not only does it take a lot more fuel to do the same job as using conventional fuels, but it also creates environmental problems of its own.

Require More Land

Combustion of biomass products require some land where they can easily be burnt. Since, it produces gases like methane in atmosphere; therefore it can be produced in those areas which are quite far from residential homes.

Thermo chemical properties of promising dendro energy species

Wide range of species were characterized for their thermo chemical properties like specific gravity, ash content, fixed carbon, calorific value, fuel value, heating value etc., and are furnished in Table 1. Depending on factors of locality, industrial and management needs, the species can be selected, adopted and promoted.

Table 1. Thermo chemical properties of species amenable for biomass power generation

S. No.	Species	Specific gravity	Volatile matter (%)	Ash content (%)	Fixed carbon (%)	Calorific value (kcal kg^{-1})	Fuel value index	Ash deformation temperature (^{o}C)	Ash fusion temperature (^{o}C)	Lignin content (%)	Heating value (MJ kg^{-1})
1	*Acacia holosericea*	0.54	66.53	4.58	21.25	3692	27.29	1043.33	1080.00	29.67	27.84
2	*Acacia auriculiformis*	0.49	66.83	3.50	21.00	3557	30.23	1080.00	1140.00	30.00	28.45

S. No.	Species	Specific gravity	Volatile matter (%)	Ash content (%)	Fixed carbon (%)	Calorific value (kcal kg^{-1})	Fuel value index	Ash deformation temperature (°C)	Ash fusion temperature (°C)	Lignin content (%)	Heating value (MJ kg^{-1})
3	*Anacardium occidentale*	0.50	66.33	4.07	21.60	3441	28.16	1100.00	1180.00	30.00	28.38
4	*Bambusa vulgaris*	0.54	66.21	3.71	22.92	4130	109.86	1130.00	1210.00	33.17	30.57
5	*Bambusa bambos*	0.56	66.66	2.92	23.08	4150	115.51	1136.67	1205.00	32.00	30.59
6	*Casuarina equisetifolia*	0.58	67.62	2.33	23.38	4593	251.65	1156.67	1230.33	34.33	31.35
7	*Cassia siamea*	0.48	65.87	4.47	21.17	3462	25.15	1050.00	1110.00	28.00	28.18
8	*Caesalpinia pulcherimma*	0.43	66.00	4.38	22.12	3479	23.32	1080.00	1120.00	28.17	28.21
9	*Dalbergia sissoo*	0.58	66.43	2.88	23.75	4314	236.81	1136.67	1230.00	33.25	30.81
10	*Eucalyptus tereticornis*	0.61	67.00	2.50	23.85	4614	240.69	1150.00	1241.67	33.67	31.02
11	*Gliricidia sepium*	0.44	65.50	4.75	20.58	3913	21.31	1083.33	1136.67	29.00	27.61
12	*Lannea coromandelica*	0.45	66.33	4.37	21.30	3812	23.29	1060.00	1136.67	27.00	27.85
13	*Leucaena leucocephala*	0.62	68.08	1.75	24.33	4552	265.68	1173.33	1273.33	34.42	31.50
14	*Prosopis juliflora*	0.78	68.17	1.48	24.88	4860	303.72	1220.00	1290.00	35.17	31.69
15	*Simarouba glauca*	0.55	65.83	4.97	21.70	3781	31.48	1056.67	1196.67	29.33	28.02

BENEFITS OF DENDRO ENERGY

- Dendro energy resources help to develop decentralized energy generation process.
- It help to convert large stretches of waste land available in the country into a productive land use system thereby, help to convert unprodctive land into a profitable productive land.
- It helps to generate income and empolyment generation activities through a massive afforestation and reforestation programme.
- Dendro biomass based enegy generation will help to connect remote in accessibel areas into a developemental process.
- It is a very cheep source of energy.

- It also helps to satisfy the timber and other non-timber requirements of the local socity besides, generating energy.
- It helps to generate electrocity where, grid connectivitiy is a problem
- Dendro biomass energy is a clean and green energy process and help to balance the oxigen and CO_2 concentration in the atmosphere.
- It help to establish different capacities of energy generation process
- It may lead to generate energy at low cost through promotion of captive power plantations.
- it reduces all emission and pollutions thereby help to achive emission reduction target envisaged by the government.
- It reduces the dependancy and burden on fossil fuels and there by help to reduce country's foreign exchequer through reduction in imports of fossil fuels.
- Above all dendro biomass energy is a renewable source of energy, clean, green, and has a potential to generate huge rural employement opportunities.

■■■■

CHAPTER - 14

GLOBAL WARMING

The most pronounced word globally in the recent past is the "global warming". The earth is continued to be warmer in the recent past due to the increased concentration of carbon-di-oxide in the atmosphere. The carbon-di-oxide acts as a one way filter and causes the temperature increase in the atmosphere. Hence there is a need to understand the process of temperature rise and measures have to be taken to control and limit the global warming process. Global warming process has been incorporated as a major subject in several graduate and post graduate programme and hence the current chapter is conceived and presented with the following sub titles.

- Definition
- Green House Effect and Global Warming
- Causes, Impacts and Effects
- Mitigation Measures
- Adaptation Measures
- Preventive Measures

An increase in the concentration of carbon dioxide in the atmosphere is one of the major causes of air pollution. Scientists believe that an increase in the amount of carbon-di-oxide and certain other gases is responsible for a phenomenon called global warming. Global warming refers to the unnatural increase in the Earth's surface temperature.

The phenomenon whereby the Earth's atmosphere traps solar radiation because of the presence of gases like carbon dioxide, water vapour and methane is called greenhouse effect. Without the natural greenhouse effect, the Earth's surface would have been too cold to sustain life. It is the increase in the amount of greenhouse gases beyond natural limits that is a cause of worry.

DEFINITION

"Global warming is an average increase in the temperature of the atmosphere near the Earth's surface and in the troposphere, which can contribute to changes in global climatic pattern".

"Global warming often refers to the warming that occur as a result of increased emissions of green house gases from human activities".

GREEN HOUSE EFFECT AND GLOBAL WARMING

The gases viz., carbon-di-oxide (CO_2), Chlorofluorocarbon (CFCs), Methane (CH_4), Ozone (O_3), Nitrous oxide (N_2O) etc. act like a glass of a green house. It absorbs the light radiation from the sun and increases the temperature of the earth. The atmosphere is a huge green house and the increase in the concentration of green house gases traps more sunlight and do not reflect them back resulting in warming of the atmosphere. This process is called green house effect which leads to global warming.

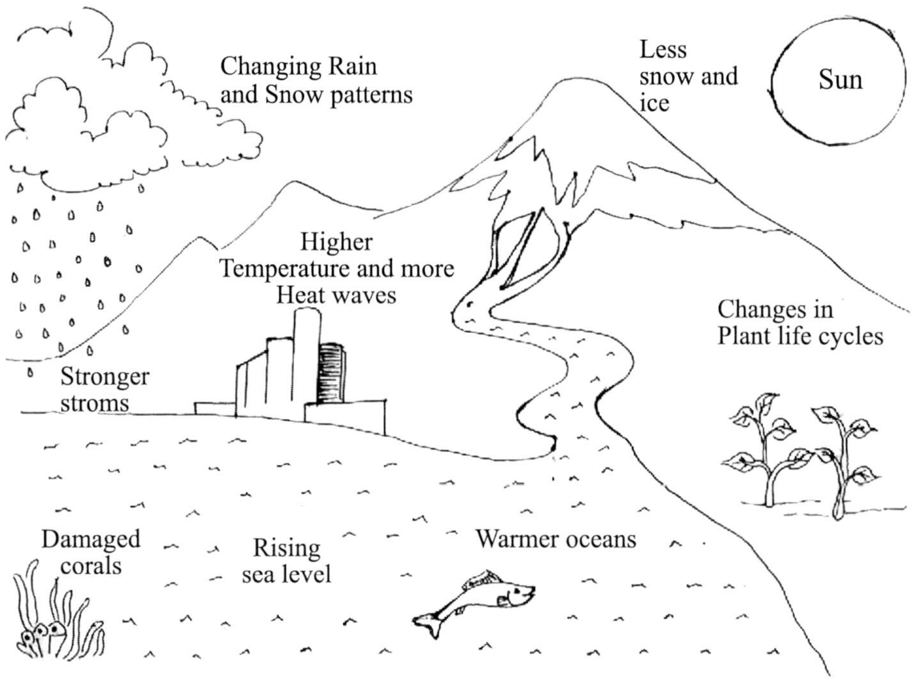

Impacts of Global Warming

GREENHOUSE GASES (GHGS) AND THEIR SOURCES

The main greenhouse gases viz, Carbon dioxide (CO_2); Methane (CH_4); Nitrous oxide (N_2O); Hydro fluorocarbons (HFCs); Per fluorocarbons (PFCs); and Sulphur hexafluoride (SF_6). The impact of any GHG is based on the magnitude of the rise in its concentration, its duration in the atmosphere and the wavelength of radiation that is absorbs.

1. **Carbon dioxide** is the GHG which is present in the largest concentration in the atmosphere. Its emission chiefly comes from fossil fuel combustion. It is showing a rise of about 0.5% per annum.

2. **Chlorofluorocarbons (CFCs)** are produced due to anthropogenic activity. Ozone is present in the stratosphere where ultraviolet (UV) radiations convert oxygen into ozone. Hence, the UV rays do not reach the Earth's surface. The CFCs which goes into the stratosphere destroys the ozone, which is evidently seen over Antarctica. The reduction of ozone concentration in the stratosphere is known as the ozone hole. This permits the UV rays to pass through the troposphere.
3. **Nitrous oxide** is naturally produced by oceans and rainforests. Man-made sources of nitrous oxide include nylon and nitric acid production, the use of fertilizers in agriculture, cars with catalytic converters and the burning of organic matter.
4. **Hydro fluorocarbons(HFCs)** are used as refrigerants, especially after the ozone-destroying CFCs had been under the Montreal Protocol.
5. **Per fluorocarbons (PFCs):** Emitted as a result of production of fluorites, they have an atmospheric lifetime of more than 1,000 years.
6. **Sulphur hexafluoride (SF6):** The most powerful greenhouse gas yet discovered, it is emitted as result of production of fluorites.

CAUSES AND IMPACTS OF GLOBAL WARMING

The global warming has the following serious impacts-

i. **Raise of sea level:** The climate change and the associated global warming occur over a long period of time. However the earth becomes warmer in recent times and it was found that 9 hottest years occurred during the last two decades. The mean earth temperature in 1997 was 0.43°C higher than the past 26 years.

ii. **Melting of polar ice:** Due to global warming and increase in temperature, the polar ice caps started melting which results in sea level increase. The increase is projected to be 18 cm by 2030 and 58 cm by 2090. This may cause flooding in low lands and many islands may submerge in the sea water.

iii. **Change of rainfall pattern:** Climate change and its associated global warming also affect rainfall and food production in agriculture. The rainfall may be highly erratic resulting in increased rainfall in certain areas and decreased rainfall in other areas which will result in flooding and drought in the respective areas. These climatic changes may cause problem in flowering, seed setting and ultimately reduced food production in developing countries may lead to hunger for many people.

iv. **Melting of the ice peaks:** Due to melting of the ice peaks, there is loss of habitat near the poles. Now the polar bears are considered to be greatly endangered by the shortening of their feeding season because of declining ice packs.

v. **Melting glaciers:** There is a significant melting of old glaciers.

vi. **Spread of disease:** There is spread of diseases like malaria due to migration to newer and currently warmer regions.

vii. **Thinning of Coral Reefs:** Due to warming seas as well as acidification because of carbonic acid formation: Almost one-third of coral reefs are now severely damaged by warming seas.

viii. **Loss of Plankton owing to warming seas:** The large (900 miles long) Aleutian island ecosystems consisting of whales, sea lions, sea urchins, kelp beds, sh, and other aquatic animals, has now reduced due to loss of plankton

MITIGATION MEASURES

- Practice Energy efficiency
- Greater use of renewable energy
- Electrification of industrial processes
- Efficient means of transport implementation: electric public transport, bicycle, shared cars
- Carbon tax and emissions markets

ADAPTATION MEASURES

In terms of adaptation measures, there are several actions that help reducing vulnerability to the consequences

- More secure facility locations and infrastructures
- Landscape restoration (natural landscape) and reforestation
- Flexible and diverse cultivation to be prepared for natural catastrophes
- Research and development on possible catastrophes, temperature behavior, etc.
- Preventive and precautionary measures (evacuation plans, health issues, etc.)

PREVENTIVE MEASURES FOR GLOBAL WARMING

As global temperatures rise, global warming is an increasing concern. The current level of carbon dioxide in the atmosphere is higher than it has been at any point in history, according to NASA, and the rise in global temperatures is moving at a faster rate than any point in the past 1,300 years. Scientific organizations, including NASA and the National Oceanic and Atmospheric Administration (NOAA) agree that human behaviors and activities are a primary cause of the warming trend. To mitigate the existing damage and prevent global warming from accelerating, the following measures can be practical.

1. Increased Afforestation

The Afforestation and reforestation activities may act as a excellent sink for reducing the CO_2 concentration in atmosphere. During photosynthesis trees absorbs more CO_2 and release O_2 there by helps to maintain balance in atmosphere. it has been estimated that single tree will absorb approximately 1 tonne of CO_2 during its life time and hence Afforestation and reforestation acts as the major strategy for reducing global warming.

2. Reduce, Reuse and Recycling

The 3R strategy is very significant in reducing the levels of green house gasses and to combat the global warming. Reducing the wastages, reusing the products coupled with recycling of the utilized materials can play a vital role in reducing the green house gas emission. it is estimated that by recycling 50% of the house hold waste approximately 2,400 pounds of CO_2 emission can be reduced.

3. Limited use of fossil fuel

The utility of fossil fuel like coal and other petroleum products are major contributor of increased green house gas emission. Hence reduced use of fossil fuel or utilization of alternative source of energy could play a vital role in reducing the green house gas emission and ultimately reduce the effect of global warming.

4. Utilization of renewable resources

The development and utilization of renewable resources like solar, wind, water, and biomass is one of the potential options to combat the environmental pollution and to reduce the global warming. The use of biomass based energy generation has been considered as the co-neutral project and is amenable for decentralized power generation strategy and hence it may be a potential strategy to reduce the global warming.

5. Promotion of biofuels

Fossil fuel as a major contributor for most of the green house gasses which necessitated searching for alternative fuels. Among the alternative fuels the potential of befoul like biopetrol (ethanol) and biodiesel is very well witnessed. Hence deployment of befouls like ethanol for petrol and biodiesel for diesel either fully or partly helps to reduce the green house gas emissions and the associated global warming.

6. Changes in the agricultural practices

Several defaulty agricultural practices are major source of emission of green house gasses. Hence change in agricultural pattern like deploying organic agriculture, efficient utilization of fertilizer and plant protection measures, No tillage/ low tillage operations, changing cropping pattern etc could play a vital role in reducing the impact of global warming.

7. Reduced deforestation

Deforestation is one of the major cause for global warming and it is estimated that the deforestation rate at the global level is over 3 million ha/annum. Several countries in Africa and Asia still witnessed wide range of deforestation. Reduction in deforestation will help to sequester more carbon and reduce the global warming process.

■■■■

CHAPTER - 15

CARBON SEQUESTRATION

The growing concerns on environmental quality are creating serious issues among the general public, planners and policy makers. The increasing concentration of carbon dioxide in the atmosphere is one of the major threats and is linked with several environmental crisis. There are several impact forecasted like changes in the weather pattern, rise in sea level, increasing temperature, changes in crop behaviour, migration of plants and animals and also threats to human health. Hence, there is a need to control carbon concentration and adopt strategies for mitigation and adaption. Among several strategies, the potential of carbon sequestration by plants in general and trees in particular has gained significant attraction across the globe. Against this backdrop, this chapter on carbon sequestration is conceived and presented with the following sub themes to benefit the readers.

- Definition
- Important Terminologies and Definitions
- Kind of Sequestration
- Methods of sequestration
- C Sequestration potential of Natural forest
    - Global Carbon Stock
    - India's Carbon stock
- C Sequestration potential of Agroforestry

DEFINITION

"Carbon sequestration is the process involved in carbon capture and long-term storage of atmospheric carbon dioxide". Carbon sequestration involves long-term storage of carbon dioxide or other forms of carbon to mitigate or defer global warming.

Carbon sequestration can be defined as the capture and secure storage of carbon that would otherwise be emitted to or remain in the atmosphere. This strategy aims

- To keep carbon emissions produced by human activities from reaching the atmosphere by capturing and diverting them to secure storage, or
- To remove carbon from the atmosphere by various means and store it.

IMPORTANT TERMINOLOGIES AND DEFINITIONS

Carbon sink

A carbon sink is a natural or artificial reservoir that accumulates and stores carbon containing chemical compound for an indefinite period. The process by which carbon sinks remove carbon dioxide (CO_2) from the atmosphere is known as carbon sequestration.

Growth of trees is mainly due to absorption of carbon dioxide from air, nutrients and water from soil and conversion of the product into biomass. About 50% of the dry weight of a tree is because of carbon. Global carbon is held in a variety of forms such as swamps, wetlands, forests and forest soils, grasslands, agriculture crops etc. which are called stocks. A stock that stores carbon is called carbon sink.

Carbon source

Carbon is a basic element for sustaining life and the main element in all classes of macromolecules. All cells require carbon as a major nutrient. On a dry weight basis, a typical cell is 50% carbon. Carbon source refers to any carbon containing molecule (carbohydrate, amino acid, fatty acid, CO_2) used by an organism for the synthesis of its organic molecules. Depending on their carbon source, organisms are either heterotrophs, requiring one or more organic compounds as their carbon source, or autotrophs, where CO_2 is the carbon source.

Carbon Stock

It refers to the absolute quantity of carbon held within a pool at a specified time.

Carbon Pool

Carbon pool is a reservoir of carbon. It is a system which has the capacity to accumulate or release carbon.

Carbon flux

As forest grows, biomass accumulates over decades and even centuries resulting in significant sequestration (storage) of carbon. This transfer of carbon from atmosphere to forest is called as carbon flux. Transfer of carbon from one carbon pool to another in units of measurement of mass per unit area and time (e.g., t C ha -1 yr-1).

Carbon neutral

It refers to a mature forest balancing sequestration of carbon dioxide and release of oxygen.

Carbon footprint

Carbon footprints refer to direct emissions (typically from energy used in home and in transport, including travel by cars, airplanes, rail and other public transport), or indirect emissions as a result of goods and services consumed). It is a measure of the impact, human activities have on the environment in terms of the amount of greenhouse gases produced, measured in units of carbon dioxide. Each unit is equivalent to reduction of one metric tons of CO_2. The term owes its origins to the idea that a footprint is what has been left behind as a result of the individual's activities. It is meant to be useful for individuals, nations and organizations to form an idea of their personal (or organizational) impact consequences in contributing to global warming.

Carbon credits

Carbon credits are of two distinct types viz., Carbon-Offset Credits (COCs) and Carbon Reduction Credits (CRCs)

Carbon-Offset Credits (COCs) means the production of clean forms of energy like wind, solar, hydro and biomass based energy generation. Carbon Reduction Credits (CRCs) means collection and storage of carbon from atmosphere through afforestation, reforestation, ocean and soil collection and storage in any other forms. Both approaches are recognized as effective ways to reduce the global carbon emission crisis.

Kinds of carbon sequestration

i. Ocean Sequestration
ii. Geologic Sequestration
iii. Terrestrial Sequestration
iv. Forests as carbon sequestration

i. Ocean Sequestration

Carbon stored in oceans through direct injection or fertilization.

ii. Geologic Sequestration

Natural pore spaces in geologic formations serve as reservoirs for long-term carbon dioxide storage. "The process of removing carbon from the atmosphere and depositing it in a reservoir". When carried out deliberately, this may also be referred to as carbon dioxide removal, which is a form of geo-engineering.

iii. Terrestrial Sequestration

A large amount of carbon is stored in soils and vegetation, which are our natural carbon sinks. Increasing carbon fixation through photosynthesis, slowing down or reducing decomposition of organic matter, and changing land use practices can enhance carbon uptake in these natural sinks.

iv. Forests as carbon sequestration

Forests in general and trees in particular are excellent sequester of carbon. Trees absorb carbon from the atmosphere and utilize them through photosynthesis and produce biomass which utilizes the plant growth and development. Due to this process the concentration of carbon-di-oxide in atmosphere is very well balanced and reduces the risks of global warming. Due to this sequestration ability, the IPCC has identified agroforestry as one of the potential land use system which has sequest huge volume of carbon and balance concentration of atmosphere.

Methods of carbon sequestration

1. Above ground biomass carbon sequestration

The above ground biomass includes plant, animals and litter from these. The proportion of carbon stored in these materials varies widely depending upon the species. In general plants fix the carbon by consumption of CO_2 through the photosynthesis mechanism. In this atmospheric carbon is converted in to plant biomass and retained within it until it decays. Woods of trees contain about 25-48 percent by carbon based on its dry weight. So forest has a huge potential in carbon sequestration.

Approaches for above ground biomass carbon sequestration

- Sustaining the forest cover
- Decreasing deforestation
- Regeneration of natural forests
- Establishing tree plantations and agroforestry
- Improving management of agricultural soils and rangelands
- Improving the fertilizer use efficiency
- Improving the diet of ruminants

Below ground carbon sequestration

Below ground carbon sequestration includes carbon fixation by soil, soil microbes, etc. Carbon stocks in soil exceed the carbon stocks in the vegetation by a factor of two to five. The global soil hold twice as much as carbon as the atmosphere (1400-1500 Gt C). The soil carbon pool comprises two components. Soil and vegetation in together exchanges 100 Gt C per year. In India, the amount of carbon stored in the soil is 23.4-27.1 Gt which is 1.6-1.8% of the global reserve.

Soil carbon sequestration

It generally refers to the medium and long-term (15-50 years) storage of C in the terrestrial soil ecosystem through various physical, chemical, biochemical and biological processes in soils. Soils have the capacity to accommodate substantial amount of C from the atmosphere by photosynthesis and sequester it for a long. The soils have an inherent upper limit or "C saturation level" above which no additional C can be stored. The process controlling saturation limit can be modified and controlled to the soil C sink and the time period over which the soil can be exploited for C sequestration.

Mechanism of soil carbon sequestration

Physical carbon sequestration

In this soil organic carbon (SOC) is held with soil mineral matrix and aggregates which inhibits microbes accessibility and their enzymes. Secondly, SOC is adsorbed in to the clay surfaces which is also not reachable by microbes. The penetration of SOC into interlayer spaces is also do the same.

Chemical protection and stabilization mechanisms

In this mechanism, the SOC is converted in to a non degradable carbon material through various chemical processes such as charcoal burning, humification, etc.

Biological protection and stabilization mechanisms

Biological organism improves the soil aggregate formation and hence stabilizes the SOC indirectly. Certain microbial products are highly resistant to the degradation. *E.g.,* Glomalin protein of VAM.

Approaches for soil carbon sequestration

- Restoration of degraded soils
- Improved management of grassland and forest soils
- Minimum or conservation tillage
- Water conservation
- Residue and manure management
- Mulching practices
- Management for reducing soil erosion

CARBON SEQUESTRATION BY FORESTS

Forests play an important role in the global carbon cycle by acting as the interface between the atmosphere and soil. Thus, forest acts as a switch between source and sink for carbon. Forests contain about 50% of the global terrestrial biomass carbon stocks (IPCC, 2007; FAO, 2000). Several studies have indicated that the global potential for enhancing carbon storage in forest and agricultural ecosystems may be as much as 60-90 Petagrams of carbon. It is believed that the goal of reducing carbon sources and increasing the carbon sinks can be achieved efficiently by protecting and conserving the carbon pools in existing forests.

(a) Global Carbon stock

Forest play an important role in climate change mitigation and adaptation process Forest sequest carbon through it photosynthetic activities and store them in their wood and soil. This sequestration potential of tree is more than any other land use system and can play a vital role in moderating the climate change process.

This sequestration potential of forest has attracted stake holders which resulted in quantification of actual carbon stock by various forest types distributed across the world. The forest in general and trees in particular capture carbon during the process of growth and development and at maturity the tree is housed with carbon.

The FAO during its 2015 assessment estimated carbon stock potential of global forest and is furnished in Table 1 & 2

It has been estimated that the world forest is housed with 485141 million tonnes of carbon which included 455893 million tonnes in actual forest resources and 29 248 million tonnes in other wooded lands. This carbon stock of world forest is contributed by various kinds of biomass as indicated below.

- Above ground biomass
- Below ground biomass
- Dead wood
- Litter
- Soil organic carbon

Among the various biomass, the above ground biomass contributed more towards carbon stock (208126 million tonnes) followed by soil organic carbon (188960 million tonnes), below ground biomass (51258 million tonnes), litter (21263 million tonnes) and dead wood (15534 million tonnes).

Similarly the carbon stock potential of forest and wood lands in each and every country has also been estimated by FAO and presented in its status report (FAO, 2015).

Only the top ten countries recorded with higher carbon stock potential alone has been identified and presented in Table 1. Among the top ten countries, the

Russian federation recorded 127900 million tonnes of carbon followed by Brazil, USA, Congo, Indonesia, Colombia, Tanzania, China, India and Guyana as indicated in Table 1

In case of carbon stock potential of wooded lands, Among several countries Argentina ranked first with an estimated carbon potential of 4114 million tonnes followed by South Africa, Russian federation, Brazil, Madagascar, USA, Mozambique, Chad, Congo and Tanzania (Table 2)

Table 1. Top Ten countries of Forest Carbon Stock

Country/Territory	Forest carbon stock (million tonnes)					Total	Percentage
	Above-ground biomass	Below-ground biomass	Dead wood	Litter	Soil		
Russian Federation	26200	6600	7450	9650	78000	127900	28.05
Brazil	48741	10481	737	1205	21065	82229	18.04
USA	14466	2864	2412	4535	16950	41227	9.04
Congo	15678	3763	-	324	9918	29683	6.51
Indonesia	9391	3098	-	-	-	12489	2.74
Colombia	7249	1617	-	-	-	8866	1.94
Tanzania	4386	1053	54	-	2506	7999	1.75
China	5144	1642	1041	-	-	7827	1.72
India	2006	641	27	132	3948	6754	1.48
Guyana	3900	909	215	93	1598	6715	1.47

(FAO, 2015)

Table 2. Top Ten countries of Woodland Carbon Stock

Country/Territory	Wooded land (million tonnes)					Total
	Above-ground biomass	Below-ground biomass	Dead wood	Litter	Soil	
Argentina	1127	361	-	180	2446	4114
South Africa	669	322	139	-	1547	2677
Russian Federation	260	130	260	-	2000	2650
Brazil	157	381	3	83	1458	2082
Madagascar	627	125	-	31	685	1468

Country/Territory	Wooded land (million tonnes)					Total
	Above-ground biomass	Below-ground biomass	Dead wood	Litter	Soil	
USA	230	43	29	447	675	1424
Mozambique	696	195	-	30	447	1368
Chad	336	134	-	41	743	1254
Congo	460	92	-	22	493	1067
Tanzania	66	16	9	-	937	1028

(FAO, 2015)

Table 3. Carbon Stock in potential of world forest and wood lands

S.No	Component	Carbon Stock in world forest	Carbon Stock in wood lands	Total
1.	Above Ground Biomass	201004	7122	208126
2.	Below Ground Biomass	48707	2551	51258
3.	Dead wood	14951	583	15534
4.	Litter	20003	1260	21263
5.	Soil Organic Carbon	171228	17732	188960
Total		**455893**	**29248**	**485141**

(FAO, 2015)

(b) India's Carbon stock

The FSI has conducted forest assessment along with carbon stock potential of Indian forest. The carbon stock potential of forest is gaining significant attraction and attention across the world particularly after 1997 kyoto protocol. The forest and climate change have an integral approach to regulate climate, rain, ground water and soil of earth. The forest acts both as a source and sink and their sink potential is very significant.

The FSI has assessed the carbon stock potential of Indian forest and state wise carbon stock potential of forest are presented in Table 4

It is estimated that Indian forest has a carbon sequestration potential of 7082 million tonnes as per 2017 assessment and this estimate is 39 million tonnes increase over the 2015 assessment. The annual rate of increase of carbon stock of Indian forest is estimated at 19.5 million tonnes which is almost equivalent to 71.5 million tonnes of carbon-di-oxide. This carbon stock assessment has been carried out by assessing the following components.

Above ground biomass : It includes all living biomass above the soil like stem, stump, branches, leaves, bark, seeds etc.

Below ground biomass : It includes the root portion upto 2mm diameter.

Dead wood : It includes all non living woody biomass either standing or lying under ground and includes root as well

Litter : It includes all non living biomass with diameter less than 5 mm.

Soil organic carbon : It includes organic carbon in mineral and organic soils upto a depth of 30 cm.

The carbon stock potential of various forest types in different states were also assessed and it is reported that Arunachal Pradesh has recorded a maximum carbon stock of 994.5 million tonnes followed by Madhya Pradesh (695.5 million tonnes), Chhattisgarh (560.9 million tonnes) and Maharashtra (493.0 million tonnes).

The per hectare carbon stock was maximum in Andaman and Nicobar islands (170.68 tonnes per hectare) followed by Arunachal Pradesh (148.52 tonnes per hectare), Sikkim (145.14 tonnes per hectare) and Karnataka (126.52 tonnes per hectare). Among the different pools of carbon stock, soil organic carbon registered 56% followed by the above ground biomass which recorded 32% of carbon stock.

Table 4. Change in carbon stock in forest between 2015 and 2017 in India (Million tonnes)

S.No	Component	Carbon Stock in forest as per ISFR 2015	Carbon stock in forest in 2017	Net change in Carbon stock	Annual increase in Carbon stock
1.	Above Ground Biomass	2220	2238	18	9.00
2.	Below Ground Biomass	695	699	4	2.00
3.	Dead wood	29	30	1	0.50
4.	Litter	131	136	5	2.50
5.	Soil Organic Carbon	3969	3979	10	5.00
6.	Total	7044	7082	38	19.00

(FSI, 2015, 2017)

Table 5. State wise Carbon stock of natural forest in India ('000' tonnes with per ha stock in tonnes)

S.No	State	AGB	BGB	Dead Wood	Litter	SOC	Total
1.	Andhra Pradesh	100539 (35.72)	38585 (13.71)	568 (0.20)	4527 (1.61)	118471 (42.09)	262690 (93.33)
2.	Arunachal Pradesh	243462 (36.36)	53378 (7.97)	4305 (0.64)	16231 (2.42)	677163 (101.12)	994,539 (148.52)
3.	Assam	47343 (16.85)	10824 (3.85)	1093 (0.39)	5240 (1.86)	112352 (39.98)	176,852 (62.93)
4.	Bihar	19063 (26.12)	6707 (9.19)	138 (0.19)	625 (0.86)	28864 (39.55)	55,397 (75.90)
5.	Chhatisgarh	206678 (37.21)	68,159 (12.27)	2588 (0.47)	7628 (1.37)	275927 (49.67)	560,980 (100.99)
6.	Delhi	230 (11.98)	52 (2.71)	2 (0.10)	11 (0.57)	653 (34.01)	948 (49.38)
7.	Goa	5153 (23.12)	1512 (6.78)	250 (1.12)	417 (1.87)	11684 (52.42)	19,016 (85.31)
8.	Gujarat	32668 (22.14)	11719 (7.94)	322 (0.22)	993 (0.67)	64995 (44.04)	110,697 (75.01)
9.	Haryana	3736 (23.53)	1269 (7.99)	20 (0.13)	74 (0.47)	7312 (46.05)	12,411 (78.15)
10.	Himachal Pradesh	70655 (46.79)	18691 (12.38)	739 (0.49)	2511 (1.66)	83186 (55.09)	175,782 (116.41)
11.	Jammu & Kashmir	112919 (48.59)	30083 (12.94)	1004 (0.43)	3529 (1.52)	128391 (55.24)	275,926 (118.72)
12.	Jharkhand	86006 (36.52)	33173 (14.08)	438 (0.19)	1298 (0.55)	101967 (43.29)	222,882 (94.63)
13.	Karnataka	128098 (34.11)	35045 (9.33)	2545 (0.68)	19745 (5.26)	289652 (77.14)	475,085 (126.52)
14.	Kerala	74166 (36.50)	19245 (9.47)	1058 (0.52)	7436 (3.66)	153976 (75.77)	255,881 (125.92)

S.No	State	AGB	BGB	Dead Wood	Litter	SOC	Total
15.	Madhya Pradesh	266040	101516	1654	7741	318713	695,664
		(34.37)	(13.11)	(0.21)	(1.00)	(41.17)	(89.86)
16.	Maharashtra	142651	48947	1986	9385	290052	493,021
		(28.15)	(9.66)	(0.39)	(1.85)	(57.23)	(97.28)
17.	Manipur	27253	8821	530	3909	102578	143,091
		(15.71)	(5.09)	(0.31)	(2.25)	(59.14)	(82.49)
18.	Meghalaya	25168	6835	881	5184	117772	155,840
		(14.68)	(3.99)	(0.51)	(3.02)	(68.69)	(90.89)
19.	Mizoram	15359	3173	633	2652	73224	95,041
		(8.45)	(1.74)	(0.35)	(1.46)	(40.26)	(52.26)
20.	Nagaland	16151	4150	666	2432	101661	125,060
		(12.93)	(3.32)	(0.53)	(1.95)	(81.40)	(100.14)
21.	Odisha	152525	50407	2108	9,087	238776	452,903
		(29.71)	(9.82)	(0.41)	(1.77)	(46.50)	(88.21)
22.	Punjab	5095	1883	26	63	8971	16,038
		(27.74)	(10.25)	(0.14)	(0.34)	(48.84)	(87.31)
23.	Rajasthan	32558	12736	216	721	43429	89,660
		(19.65)	(7.69)	(0.13)	(0.44)	(26.21)	(54.10)
24.	Sikkim	13379	3735	211	585	30624	48,534
		(40.01)	(11.17)	(0.63)	(1.75)	(91.58)	(145.14)
25.	Tamil Nadu	84067	29252	1006	5579	109434	229,338
		(31.99)	(11.13)	(0.38)	(2.12)	(41.64)	(87.26)
26.	Telangana	72498	28388	333	3117	80639	184,975
		(35.51)	(13.90)	(0.16)	(1.53)	(39.49)	(90.59)
27.	Tripura	15674	3224	556	1613	42341	63,408
		(20.29)	(4.17)	(0.72)	(2.09)	(54.80)	(82.07)
28.	Uttar Pradesh	47752	14264	444	1824	60850	125,134
		(32.53)	(9.72)	(0.30)	(1.24)	(41.45)	(85.25)
29.	Uttarakhand	105173	26961	1316	5665	145549	284,664
		(43.29)	(11.10)	(0.54)	(2.33)	(59.91)	(117.17)

S.No	State	AGB	BGB	Dead Wood	Litter	SOC	Total
30.	West Bengal	45382	13916	434	2585	100884	163,201
		(26.94)	(8.26)	(0.26)	(1.53)	(59.88)	(96.87)
31	Andaman & Nicobar Islands	39426	11901	2048	3702	57996	115,073
		(58.48)	(17.65)	(3.04)	(5.49)	(86.02)	(170.68)
32	Chandigarh	61	19	0	2	122	204
		(27.73)	(8.64)	(0.00)	(0.91)	(55.45)	(92.73)
33	Dadra & Nagar Haveli	447	106	10	35	827	1,425
		(21.59)	(5.12)	(0.48)	(1.69)	(39.95)	(68.84)
34	Daman & Diu	11	2	0	2	76	91
		(5.50)	(1.00)	(0.00)	(1.00)	(38.00)	(45.50)
35	Lakshdweep	55	*0	1	5	100	161
		(20.37)	(0.00)	(0.37)	(1.85)	(37.04)	(59.63)
36	Puducherry	108	23	1	8	311	451
		(20.00)	(4.26)	(0.19)	(1.48)	(57.59)	(83.52)
Total		**2237549**	**698701**	**30130**	**136161**	**3979522**	**7,082,063**
		(31.59)	**(9.86)**	**(0.43)**	**(1.92)**	**(56.19)**	**(99.99)**

(FSI, 2017)

Table 6. Forest Type based Carbon stock in india ('000' tonnes with per ha stock in tonnes)

S.No	Forest Type	AGB	BGB	Dead Wood	Litter	SOC	Total
1.	Tropical Wet Evergreen Forests	100004	34458	3567	9216	180825	328070
		(47.91)	(16.51)	(1.71)	(4.42)	(86.64)	(157.18)
2.	Tropical Semi-Evergreen Forests	274741	56508	5861	32009	755307	1124426
		(28.02)	(5.76)	(0.60)	(3.26)	(77.04)	(114.69)
3.	Tropical Moist Deciduous Forests	331701	68224	10752	40400	871939	1323016
		(22.92)	(4.71)	(0.74)	(2.79)	(60.24)	(91.40)
4.	Littoral and Swamp Forests	26955	9324	14	475	36796	73564
		(40.59)	(14.04)	(0.02)	(0.72)	(55.42)	(110.79)

S.No	Forest Type	AGB	BGB	Dead Wood	Litter	SOC	Total
5.	Tropical Dry Deciduous Forests	1042208 (37.29)	409233 (14.64)	4638 (0.17)	28372 (1.02)	1185892 (42.43)	2670343 (95.54)
6.	Tropical Thorn Forests	9280 (6.93)	3619 (2.70)	220 (0.16)	1263 (0.94)	30102 (22.47)	44484 (33.20)
7.	Tropical Dry Evergreen Forests	2785 (31.76)	1094 (12.47)	40 (0.46)	130 (1.480)	4198 (47.87)	8247 (94.04)
8.	Subtropical Broad leaved Hill Forest	39623 (20.68)	14600 (7.62)	672 (0.35)	4391 (2.29)	156353 (81.61)	215639 (112.55)
9.	Subtropical Pine Forests	56838 (28.57)	14347 (7.21)	769 (0.39)	2320 (1.17)	123087 (61.87)	197361 (99.21)
10.	Subtropical Dry Evergreen Forest	633 (34.59)	248 (13.55)	5 (0.27)	7 (0.38)	1271 (69.45)	2164 (118.25)
11.	Montane Wet Temperate Forests	13432 (26.94)	3391 (6.80)	338 (0.68)	1008 (2.02)	37948 (76.12)	56117 (112.57)
12.	Himalayan Moist Temperate Forest	155932 (53.37)	39361 (13.47)	1793 (0.61)	7407 (2.54)	210647 (72.10)	415140 (142.09)
13.	Himalayan Dry Temperate Forests	34057 (65.71)	9234 (17.82)	277 (0.53)	696 (1.34)	34582 (66.72)	78846 (152.12)
14.	Sub-Alpine Forests	57564 (44.54)	15607 (12.08)	647 (0.50)	1384 (1.07)	88973 (68.85)	164175 (127.04)
15.	Moist Alpine Scrubs	3092 (27.10)	838 (7.34)	51 (0.45)	88 (0.77)	7163 (62.78)	11232 (98.44)
16.	Dry Alpine Scrub	5846 (35.95)	1585 (9.75)	59 (0.36)	175 (1.08)	8064 (49.59)	15729 (96.73)
17.	Plantation/TOF	82854 (16.61)	17031 (3.41)	429 (0.09)	6820 (1.37)	246374 (49.39)	353508 (70.87)
Total		**2237545** **(31.59)**	**698702** **(9.86)**	**30132** **(0.43)**	**136161** **(1.92)**	**3979521** **(56.19)**	**7082061** **(99.99)**

(FSI, 2017)

CARBON SEQUESTRATION BY AGROFORESTRY

Agroforestry is a sustainable land use management system which increases the overall productivity by combining trees, annuals and or animals in the same land use system. This agroforestry system offers unique advantage of increasing the livelihood opportunity besides acting as a sink for carbon. Agroforestry has been identified as one of the potential land use system for sequesting carbon and acting as a source of mitigation and adaptation strategies. The carbon sequestration potential of agroforestry has been documented by plethora of researchers and the salient features are furnished.

- Agroforestry has significant carbon sequestration potential both in above and below ground levels. In the above ground biomass the carbon is stored in the standing biomass. The carbon in the below ground is stored in the form of enhancement of soil carbon and root biomass.
- The carbon present in the above and below ground biomass in an agroforestry system is generally much higher than the equivalent land use without trees.
- The agroforestry systems have the mitigation potential of 1.1 to 2.2 Pg of carbon in terrestrial ecosystem over the next 50 years.
- The conversion of 630 million ha of unproductive crop land into agroforestry will have the potential of 5,86,000 Mg of carbon per year in the year 2040.
- In India, it has been estimated that the agroforestry based land use system has the potential of sequesting 68 to 228 Mg carbon per ha (Dixon *et al.*, 1994)
- It is also estimated that the agroforestry systems could store carbon about 83.6 tonnes of carbon per ha up to 30 cm of soil depth (Jha *et al.*, 2001)
- Besides environmental benefits, agroforestry has the potential of economic benefits of increasing the productivity and farm income.
- Agroforestry also shows significant carbon accumulation in living biomass as well as in soil.
- The agroforestry systems can also contribute towards reducing carbon-di-oxide emissions.
- The agroforestry has the ability to enhance the resilience of the system and manages with the adverse impact of climate change.
- Above all agroforestry systems help to conserve soil and water and ameliorate the microclimate of the region.

■■■■

CHAPTER - 16

FOREST BASED INDUSTRIES

Forest plays a vital role in extending raw material resources to the wood based industries till the recent past. But due to policy changes and legal regulations, the industries have been directed to generate their own raw material resources without depending on forest department supply. This has created a serious concern in the raw material availability and the significance of forest and the dependent industry has been realized as a growing concern. Hence, there is a need to assess the availability of various wood based industries, status of raw materials and the associated supply chain both within and outside the country. Wide ranges of industries are prevalent and only the organized sectors of industries are recognized and the existing demand supply pattern is based on these industries only. There are wide range of unorganized industries and utilizes wood as a raw material and these industrial demand for raw materials is not very well witnessed. Hence the current chapter on Forest based industries is presented with the following themes.

- Major wood based industries
- Wood demand for major industries
- Wood production
- Wood consumption
- Raw material and species

WOOD BASED INDUSTRIES

Forests in India, are primarily treated as social and environmental resource and secondarily as a commercial resource. The country's demand for wood and wood products largely depends on private lands in the recent past. Prior to the enactment of Forest Conservation Act and National Forest Policy 1988, most wood based industries in the country were deriving their raw material predominantly from forests and continue to withdraw forest produce at concessional prices.

Some state governments entered into long term agreement with industries to supply raw material from forest without sustainable and progressive plan for raw material security. This has resulted in continued withdrawal of raw material from forests over a long span of time in unsustainable way which resulted in large scale degradation of natural forest.

After strong policy and legal restrictions these industries have been compelled to generate their own raw material resources. The National Forest Policy of 1988 for the first time in the country have provided guidelines and regulations for wood based industries and directed these wood based industries to generate their own raw material resources.

The practice of supplying raw material regularly to the industries on administered price was almost stopped. The policy guideline has directed the wood based industries to establish their own plantations through direct relationship with farmers and individuals by providing necessary inputs, technology, credit facilities and assuring price supportive mechanism.

However, many industries although a few have not taken the new policy seriously but several industries have entered into industrial wood plantation programme in association with various stake holders.

MAJOR WOOD BASED INDUSTRIES

Industries which predominantly use wood and wood residues are considered to be forest or wood based industries. The wood based industries that depend on forest or plantation for raw materials are pulp and paper, match, sawn wood, veneer, plywood, particle board, sports goods, agricultural implements, construction industries etc. The various wood based industries available in India are listed below.

1. Paper and Paperboard
2. Newsprint
3. Construction Industry
4. Furniture
5. Packaging
6. Rayon Grade Pulp
7. Automobiles
8. Agricultural implements
9. Railway Sleepers
10. Sports goods
11. Handicrafts
12. Plywood
13. Veneer
14. Particle Board
15. MDF Board

16. Match Box
17. Mining
18. Catamaram
19. Pencil Industry
20. Miscellaneous Industry

The demand for wood for these industries are increasing at an alarming rate but there is no concomitant development of plantations which resulted in massive imports.

WOOD DEMAND FOR MAJOR INDUSTRIES

The total industrial wood demand for the year 2020 has been projected to 153 million m^3 which will increase from 58 million m^3 in the year 2000.

Growth pattern of future demand of wood

Year	Demand (million m^3)	Percentage increase (Average per annum)
2010	95.00	5.68
2015	123.00	5.89
2020	153.00	4.88

The demand of wood has been categorized into short rotation species and long rotation species. The short rotation species are primarily used for pulp and paper industries, packing agricultural implements, sports goods, plywood, match and other miscellaneous industries. In most cases, the short rotation species are primarily softwood and used in most of the growing wood based industries. The MoEF in association with FAO have conducted elaborate study and projected the demand for short rotation species suitable for various wood based industries. The projected demand for short rotation species ranged between 0.35 million m^3 (veneer) and 35.84 million m^3 (paper and paper board). The overall projected demand for various wood based industries is projected at 6.76 million m^3 for the year 2015 and 87.70 million m^3 for the year 2020 respectively.

Projected demand for RWE from SR species (million m^3)

S.No	Industry	2010	2015	2020
1	Paper and paperboard (100%)	15.50	26.64	35.84
2	Newsprint (100%)	3.42	4.63	6.22
3	Rayon grade pulp (100%)	3.10	3.40	3.80
4	Construction industry (20%)	4.42	5.26	5.70

S.No	Industry	2010	2015	2020
5	Packaging (50%)	3.20	3.78	4.50
6	Agricultural implements (50%)	1.25	1.25	1.25
7	Sports goods (50%)	0.49	0.84	1.37
8	Plywood (50%)	8.98	11.45	14.60
9	Veneer (50%)	0.22	0.27	0.35
10	Matchbox (100%)	3.00	3.40	4.00
11	Mining (50%)	2.00	2.25	2.50
12	Miscellaneous industry (50%)	4.70	5.60	7.58
	Total	**50.18**	**68.76**	**87.70**

The demand for wood from long rotation species is almost equal to that from short rotation species upto the year 2005. But during 2020 the projected demand for wood from short rotation species will be 33% higher than that from long rotation species. The projected demand from long rotation species for the year 2020 ranged between 0.16 million m^3 (Catamaram) and 22.80 (construction industry). The overall projected demand for wood from long rotation species will be 54.40 million m^3 (2015) and 65.10 million m^3 (2020) respectively which are lower than the projected demand from short rotation species. The comparative analysis of demand projection of industrial wood both from short rotation and long rotation species.

Projected demand for RWE from LR species (in million m^3)

S.No	Industry	2010	2015	2020
1	Construction industry (80%)	17.68	21.04	22.80
2	Packaging (5%)	3.20	3.78	4.50
3	Furniture (100%)	4.62	5.9	7.53
4	Automobile (100%)	0.41	0.60	0.87
5	Agricultural implements (50%)	1.25	1.25	1.25
6	Railway sleepers (100%)	0.22	0.02	0.02
7	Sports goods (50%)	0.49	0.84	1.37
8	Handicrafts (100%)	0.65	0.78	0.95
9	Plywood (50%)	8.98	11.45	14.60
10	Veneer (50%)	0.22	0.27	0.35
11	Particle board (100%)	0.22	0.28	0.35
12	MDF board (100%)	0.21	0.24	0.28

S.No	Industry	2010	2015	2020
13	Mining (50%)	2.00	2.25	2.50
14	Catamaran (100%)	0.07	0.11	0.16
15	Miscellaneous industry (50%)	4.70	5.60	7.58
	Total	**44.92**	**54.40**	**65.10**

Summary demand project of industrial wood from SR & LR species (in million m^3)

S.No	Industry	2010	2015	2020
1	Wood from SR species	50.18	68.76	87.70
2	Wood from LR species	44.92	54.40	65.10
	Total (% growth)	95.10 (5.68)	123.16 (5.89)	152.80 (4.88)

(FAO, 2009)

WOOD PRODUCTION

The growing stock of Tree Outside Forest has remained almost static across recent decades, so it can be assumed that timber production from Tree Outside Forest has also remained static or grown marginally—considering a marginal growth of 25 million cum in the 2015 assessment over the 2011 assessment (TOF growing stock in 2015 was 1,573 million cum).

According to the FAO's 2015 report on Global Forest Assessment, the total wood extracted in India in 2015 was 434.77 million cum, out of which the fuel wood volume was 385.25 million cum or 231.15 million tonnes.

Annual Estimated Wood production in India

Categories	Production (million m^3)
Estimated production of timber from forests (Excluding FDCs)	1.205
Timber production from Forest Development Corporation (FDCs)	1.97
Annual production of timber from TOF	44.34
Bamboo production in India	5.38
Imports (All timbers and allied products in 2015)	18.01
Total Timber production	70.9
Fuel wood production	385.25
Total wood production	456.15

Annual Wood Consumption in India

Categories	Wood Consumption (million m³)
Construction, furniture and agriculture implements	48.00
Plywood and Panels	8.47
Paper, Paper board and newsprint	12.52
Fuel wood	333.00
Total Annual wood consumption	402.00
Total Annual Timber consumption	69.00

(CSE, 2017)

RAW MATERIALS AND SPECIES

Many industries are unique to certain species but industries like pulp and paper, plywood, veneer, match, MDF and construction industry use multifarious species. Many industries for want of suitable raw material started utilizing the alternate species available. The choice of species for various wood based industries has been characterized as short, medium and long rotation.

Rotation

A rotation period is a planned number of years or the time period between the establishment of a stand of trees and to the harvestable stage of the same stand or the final felling.

Rotation of tree species are categorized as,

i. Short Rotation
ii. Medium Rotation
iii. Long Rotation

i. Short Rotation

Short rotation is the term refers to species that are fast growing and are harvested in a very short period of time i.e., within 5 - 6 years.

Examples: *Casuarina* spp., *Eucalyptus* spp., *Melia dubia, Neolamarckia cadamba.*

ii. Medium Rotation

Medium rotation species are those that are harvested between 7 to 15 years where the species acquires the harvestable property at that age.

Examples: *Ailanthus excelsa*, *Dabergia sissoo, Swietenia macrophylla, Acrocarpus fraxinifolius.*

iii. Long Rotation

Long rotation crops have the rotation age of more than 15 years, where the species is harvested only after 20 years of age.

Examples: *Tectona grandis, Dalbergia latifolia, Pterocarpus santalinus, Santalum album.*

SPECIES SUITABLE FOR DIFFERENT WOOD BASED INDUSTRIES

S.No.	Wood based Industries	Suitable Species
1	Paper and Paperboard	*Eucalyptus terticornis, Eucalyptus hybrid, Casuarina equisetifolia, Leucaena leucocephala,* Bamboo spp., *Acacia auriculiformis, Pinus* spp. and *Melia dubia*
2	Newsprint	*Eucalyptus camaldulensis*, Bamboo spp., *Boswellia serrata* and *Kydia calycina*
3	Construction Industry	*Acacia nilotica, Albizia* spp., *Adina cerdifolia, Artocarpus* spp, *Anoguissus* spp., *Abies pindrow, Pinus roxburghii* and *Toona ciliata*
4	Furniture	*Tectona grandis, Dalbergia sissoo, Artocarpus* spp., *Ougeinia oojeinensis, Dalbergis latifolia, Pterocarpus marsupium, Gmelina arborea* and *Shorea robusta*
5	Packaging	*Abies pindrow, Picea smithiana, Pinus wallichiana, Bombax ceiba, Mangifora indica, Neolanarckia cadamba, Ailanthus malabarica, Alstonia scholaris, Vateria indica, Boswellia serrata* and *Ailanthus excelsa.*
6	Rayon Grade Pulp	*Neolanarckia cadamba, Alstonia scholaris, Vateria indica, Eucalyptus* spp and Poplars.
7	Automobiles	*Guaiacum officinale, Guaiacum sanctum, Nanika littoralis, Hardwicka binnata, Mesua fora* and *Xylia xycocarpa*
8	Agricultural implements	*Acacia nilotica, Acacia catechu, Anogeissus latifolia, Cassia fistula, Chloroxylon swietenia, Diospyros melanoxylon, Grewia tiltifolia, Hopea oderata, Mesua ferra* and *Xylia xycocarpa*
9	Railway Sleepers	*Shorea robusta, Cedrus deodara, Mesua ferra, Xylia xylocarpa, Lagerstroemia parviflora, Pterocarpus marsupium* and *Hopea parviflora*

S.No.	Wood based Industries	Suitable Species
10	Sports goods	*Diospyros melanoxylon*, *Taxus baccata*, *Salix alba*, *Morus alba*, *Arundinaria falcate*, *Ocotea rodioei*, *Carya* spp., *Dalbergia sissoo*, *Fraxinus* spp., *Acer* spp, *Fagus* spp., *Lagerstroemia lanceolata.*
11	Handicrafts	*Adina cordifolia*, *Artocarpus beterophyllus*, *Dysoxylum malabarium*, *Gardenia latifolia*, *Miliasa tomentosa*, *Mitragyna parvifolia*, *Wringhtia tinctoria*, *Gmelina arborea* and *Aegle marmelos*
12	Plywood	*Dipterocarpus macrocarpus*, *Acrocarpus fraxinifolius*, *Mangifera indica*, *Vateria indica*, *Ailanthus* spp., *Toona ciliata*, *Melia dubia*, *Eucalyptus* spp. and *Swietenia* spp.
13	Veneer	*Syzygium cumini*, *Terminalia bellerica*, *Dalbergia latifolia*, *Dysoxylum malabaricum, Swietenia* spp. and *Tectona grandis*
14	Particle Board	*Anacardium occidentle, Mangifera indica*
15	MDF Board	*Mangifera indica*, *Abies pindrow*, *Pinus wallichiana*, *Bombax ceiba*, *Neolanarckia cadamba*
16	Match Box	*Populus deltoides*, *Populus tremula*, *Tilia japonica*, *Bombax ceiba*, *Ailanthus excelsa, Ailanthus triphysa, Pinus lambertiana, Canarium strictum, Sterculia villosa, Trewia nudifolia, Neolanarckia cadamba Alstonia scholaris, Albizia falcatoria* and *Erythina* spp.
17	Mining	*Xylia dolabriformis*, *Tectona grandis*, *Hardwickia binata*, *Xylia xylocarpa*, *Acacia nilotica*, *Mesua ferrea* and *Hopea parviflora*
18	Catamaram	*Albizia falcatoria* and *Albizia lebbeck*
19	Pencil Industry	*Juniperus virginiana*, *Juniperus procera*, *Alnus nepalensis*, *Juniperus macropoda*, *Neolanarckia cadamba* and *Cedrus deodara*
20	Miscellaneous Industry (Motor Truck bodies)	*Artocarpus hirsatus*, *Calophyllum* spp, *Terminalia bialata*, *Dysoxylum malabaricum* and *Logerstroemia lanceolata*

CHAPTER 17

ORGANIZATIONS IN FORESTRY

The forest development and the associated strategies for conservation and management are carried out by both public and private sector organizations. These organizations are involved in conservation programmes, developmental process, management issues and also certification process. Wide ranges of organizations are available across the world and their attributes and function are of significant interest. The graduating professionals across the disciplines should be aware of these organizations and hence a chapter on Organizations in Forestry have been conceived and presented with the following sub title.

National Organizations	FSI, ICFRE, ANFPDC, AWBI, BSI, CAFRI, CZA, CSE, CRIDA FRI, ICAR, IPIRTI, IGNFA, KFRI, NBA, NBW, NTCA, WCCB, ZSI, WTI
International Organizations	CIFOR, ITTO, World Agroforestry Centre (ICRAF), IUCN, WWF, IUFRO, INBAR, FSC, FAO, IPCC

NATIONAL ORGANIZATIONS

1. Forest Survey of India

Forest Survey of India (FSI), is a premier national organization under the Ministry of Environment and Forests, responsible for assessment and monitoring of the forest resources of the country regularly. It is also engaged in providing the services of training, research and extension. Established on June 1,1981, the Forest Survey of India succeeded the "Preinvestment Survey of Forest Resources" (PISFR), a project initiated in 1965 by Government of India with the sponsorship of FAO and UNDP.

Objectives and Activities of FSI

- Preparation of State Forest Report once in every two years and provides assessment of latest forest cover in the country and maintains the forest cover changes.

- FSI inventorizes forest and non forest area and develop database on forest tree resources. The FSI develops thematic map on 1:50000 scale.
- It acts as a nodal agency for collection, compilation, storage and dissemination of spatial database on forest resources.
- It extends capacity building programme to forestry personnel towards application of technology related to resource survey, Remote sensing, GIS etc.
- FSI strengthens Research and Development infrastructure and carries out research on applied forest survey techniques.
- FSI supports the forest department of States and Union territories in forest resource survey, mapping and inventory.
- It also undertakes forestry related studies and consultancies.
- It carries out custom made training courses for state forest departments and other organizations on project mode.

2. Indian Council of Forestry Research and Education (ICFRE)

ICFRE was established in the year 1986 with the mandated objective of catering the needs of Forestry Research, Education and Extension. The headquarters of ICFRE is in Dehradun and it has nine regional research institutes and five centres in different biogeographical regions of the country. The regional research institutes are located at Jodhpur, Dehradun, Shimla, Hyderabad, Coimbatore, Ranchi, Bengaluru, Jorhat and Jabalpur, and the centres are at Agartala, Aizawl, Allahabad, Chhindwara and Visakhapatnam.

Objectives and Activities

- ICFRE undertake aid, promote and coordinate forestry research, education and extension.
- ICFRE develops and maintain National Information Centre for forestry and allied science.
- It acts as a clearing house for research relating to forests, biodiversity, environment and ecology.
- ICFRE involves in massive extension programme through systematic extension machinery.
- It offers consultancy services and training programmes in the field of forestry.
- ICFRE helps to generate, preserve, disseminate and advance knowledge, technologies and solutions for addressing the issues related to forest.
- ICFRE promotes linkages and interactions between people, forest and environment on a sustainable basis.

3. Andaman & Nicobar Islands Forest and Plantation Development Corporation

Andaman and Nicobar Islands Integrated Development Corporation Limited (ANIIDCO) was incorporated on 28th June 1988 under the Companies Act 1956 for rapid economic growth of the Islands. The main objective of the Corporation is to develop and commercially exploit the natural resources for the balanced and environment friendly development of the territory.

Role and objectives

- Trading of iron and steel, petroleum products, liquor and milk.
- Megapode Resort, Megapode Camping Resort, and air-ticketing agency
- Financial assistance to industrial concerns.
- Infrastructure Development: Tourism and fisheries.

4. Animal Welfare Board of India , Chennai

The Animal Welfare Board of India is a statutory advisory body on Animal Welfare Laws and promotes animal welfare in the country. Established in 1962 under Section 4 of the Prevention of Cruelty to Animals Act, 1960 (No. 59 of 1960), the Animal Welfare Board of India was started under the stewardship of Late Smt. Rukmini Devi Arundale, well known humanitarian.

Objectives and Activities

- Animal Welfare Board (AWB) keeps a law in force for the prevention of Cruelty to Animals under constant study
- It also advises the government on the amendments of laws related to animals from time to time.
- The AWB addresses the government towards making rules under the act with a view to prevent unnecessary pain or suffering to animals.
- It also advises to design the vehicles towards reducing the burden on draught animals.
- It also advises, design and establishment of slaughter houses.
- It also extends the financial assistance for the establishment of Pinjarapoles, rescue homes, animal shelters, sanctuaries and also the formation of animal welfare organizations.
- It also establishes and coordinates the work with other association and bodies towards preventing cruelty to animals and birds.
- It advises government on matter relating to medical care in animal hospitals and extends financial assistance.

- It also educates the people towards promotion of animal welfare.
- Finally, it advises the government on matter connected with animal welfare.

Botanical Survey of India (BSI)

The Botanical Survey of India (BSI) was established in 1890 with the objectives of exploring the plant resources of the country and identifying plant species with economic virtue. In 1954, the Government reorganised the BSI with the following objectives.

Role and Objectives

- BSI undertakes intensive floristic surveys and collect accurate and detailed information on the occurrence, distribution, ecology and economic utility of plants in the country.
- It collects, identify and distribute materials which may be of use to educational and research institutions.
- It acts as a custodian of authentic collections in well planned herbaria and to document the plant resources in the form of local, district, state and national flora.

5. Central Agroforestry Research Institute (CAFRI)

Originally this institute was nomenclatured as National Research Centre for Agroforestry to strengthen and coordinate the agroforestry research. It was established on May 8th, 1988 at Jhansi (U.P.). After 26 years of succcssful research and development activities on agroforestry, ICAR elevated the status of the National Research Centre for Agroforestry (NRCAF) to ICAR- Central Agroforestry Research Institute (CAFRI) from 1st December, 2014.

Role

- The main role of the institute is to improve quality of life of rural people through integration of perennials on agriculture landscape for economic, environmental and social benefits.
- It also aims towards Integration of woody perennials in the farming system to improve land productivity through conservation of soils, nutrients and biodiversity to augment natural resource conservation, restoration of ecological balance, alleviation of poverty and to mitigate risks of weather vagaries.

Objectives

- Develop sustainable agroforestry practices for farms, marginal land and wastelands in different agroclimatic zones of India.

- Coordinate network research for identifying agroforestry technologies for inter-region.
- Training in agroforestry research for ecosystem analysis.
- Transfer of agroforestry technology in various agro climatic zones.

6. Central Zoo Authority

The functioning of zoos in India is regulated by an autonomous statutory body called Central Zoo Authority which has been constituted under the Provisions of Wild Life (Protection) Act. The Authority consists of a Chairman, ten members and a Member Secretary.

Role

- Complement and strengthen the national efforts in conservation of the biodiversity of the country, particularly the fauna through the ex-situ conservation linked with in-situ practices.
- Provide better upkeep and veterinary care to the wild animal housed in zoos in India to ensure their conservation through best practices of management and bringing education & awareness among the people.

Objectives and Activities

- Specify the minimum standards for housing, upkeep and veterinary care of the animals kept in a zoo;
- Evaluate and assess the functioning of zoos with respect tot he standards or the norms as may be prescribed;
- Recognise or derecognise zoos;
- Identify endangered species of wild animals for purposes of captive breeding and assigning responsibility in this regard to a zoo;
- Co-ordinate the acquisition, exchange and loaning of animals for breeding purposes;
- Ensure maintenance of stud-books of endangered species of wild animals bred in captivity;
- Identify priorities and themes with regard to display of captive animals in a zoo;
- Co-ordinate training of zoo personnel in India and outside India and research in captive breeding and educational programmes for the purposes of zoo;
- Provide technical and other assistance to zoos for their proper management and development on scientific lines;

7. Centre for Science and Environment (CSE)

The Centre for Science and Environment (CSE) is a public interest research and advocacy organisation based in New Delhi. CSE researches into, lobbies for and communicates the urgency of development that is both sustainable and equitable.

Role and Objectives

- Raise these concerns, participate in seeking answers and – more importantly – in pushing for answers and transforming these into policy and so practice.
- To improve through research and by communicating our understanding through publications.

8. Central Research Institute for Dry land Agriculture (CRIDA)

Central Research Institute for Dryland Agriculture (CRIDA) is a National Research Institute under the Indian Council of Agricultural Research (ICAR) established in 1985 and located in Hyderabad, Telengana state

Role and Objectives

- Basic and applied research for sustainable and climate resilient agriculture in rainfed areas.
- Co-ordinate network research for generating location-specific technologies in rainfed areas.
- Centre for capacity enhancement in natural resource management in drylands.

Activities

- CRIDA has a mandate to carry out basic and applied research in rainfed farming.
- The Institute also undertakes National/ International Collaborations and Consultancy Projects.
- This is the lead Institute and the National Nodal point for the National Innovations in Climate Resilient Agriculture (NICRA) which is being implemented at large number of Research Institutes of ICAR, State Agricultural Universities and 100 KVKs.

9. Forest Research Institute (FRI)

The Forest Research Institute (FRI) is an institute of the Indian Council of Forestry Research and Education and is a premier institution in the field of forestry research in India. It is located at Dehradun in Uttarakhand, and is one of the oldest institutions of its kind. In 1991, it was declared a deemed university by the University Grants Commission.

It was founded in 1878 as the British Imperial Forest School by Dietrich Brandis. In 1906, it was reestablished as the Imperial Forest Research Institute, under the British Imperial Forestry Service. The Indian Forest College was established in 1938; officers recruited to the Superior Forest Service by the states and provinces were trained there.

Role and Objectives

The main thrust area of current research is to develop technologies for:

- Enhancement of forest Productivity
- Improvement of planting Stock
- Rehabilitation of Wastelands
- Efficient utilization of wood and non-wood forest products
- Development of eco-friendly products and processes.

10. Indian Council of Agricultural Research (ICAR)

The Indian Council of Agricultural Research (ICAR) is an autonomous body responsible for co-ordinating agricultural education and research in India. It reports to the Department of Agricultural Research and Education, Ministry of Agriculture. The Union Minister of Agriculture serves as its president. It is the largest network of agricultural research and education institutes in the world.

Objectives and activities

- Coordination, aiding and supporting National Agricultural Research System to enhance the productivity and profitability.
- Strengthening and aiding agriculture and allied sciences educational activities towards creating professional manpower needs of the country.
- Demonstrates and delivers technologies through organized extension machineries to effectively transfer the technologies to farm lands.
- Developing technologies to cater to the needs of climate smart agriculture.

11. Indian Plywood Industries Research and Training Institute (IPIRTI)

This institute was established in the year 1962 at Bangalore, Karnataka as a cooperative research association of the plywood industries. It has become an autonomous organization under the MOEF in the year 1990. This institute is a reputed organization and creates centre of expertise in the mechanical wood industries and updated with modern infrastructure.

Role and objectives

- Research on all aspects of production of sawn timber, manufacturing plywood and other allied engineered and reconstituted wood and lignocellulosic products, including improvement of materials, manufacturing processes, machines and appliances, time and motion studies – conditions of work standards in factories.
- Training in connection with forest product utilization for plywood industry and trade and allied industries. Imparting technical education and/or training at undergraduate, postgraduate, and/or any other level in technology of agro and forests products, adhesives and laminates, and/or synthetic finishing, manufacturing machinery.
- Inspection, Testing and Certification of all forest products viz. plywood, wood, timber, hardboard, particleboard, chipboard, furniture, glue-lam, compreg, doors, panel doors, block board, flush doors, veneered panels, veneers, laminated panels, composite boards, and the products of allied trade and industry.
- Extension includes transfer of technology for commercialization, entrepreneur development, information dissemination through research/ technical reports, quarterly newsletter, participation in exhibition, seminars, conferences, workshop, Scientists visit to the industry to assist in process and product development.

12. Indira Gandhi National Forest Academy (IGNFA)

Indira Gandhi National Forest Academy (IGNFA) was constituted in the year 1987 by renaming the erstwhile Indian Forest College, which was originally established in 1938 for training senior forest officers. It is situated in the New Forest campus of Forest Research Institute (FRI) on Chakrata Road (NH-72), Dehradun town. IGNFA is currently functioning as a Staff College for the officers of the Indian Forest Service (IFS).

The primary mandate of the Academy is to impart knowledge and skills to the professional foresters and help them to develop competence for managing the country forest and wildlife resources on a sustainable basis. In the Academy training is provided at different levels of seniority in the Indian Forest Service besides training the new entrants to the service.

Role and Objectives

- The IGNFA conducts Professional Forestry training for IFS Probationers through a series of class room sessions, tours, excursions and specialized modules. It nurtures young foresters capable of making difference in the management of ecological assets of our country besides inculcating human values and professional ethics amongst them.

- It involves in skill upgradation Programme for State Forest Service (SFS) officers inducted into IFS to sensitize and orient them to the ethos and functioning of All India Services and to keep them abreast with the rapidly changing scenario of forest management, development and administration.
- It facilitates and conducts Mid Career Training (MCT) programme for IFS officers in three phases of the service viz III, IV and V meant for the seniority 7 to 9, 16 to 18 and 26 to 28 years respectively. It aims to provide best training opportunities to the officers by roping in some of the leading training institutions in the country and abroad. The officers get a chance to interact with the experts of respective fields and also get exposure to best management practices of natural resources in the foreign countries.
- It offers training programmes and workshops for the three All India Services, members of Higher Judiciary, exposure to Officer Trainees of other services with an endeavor to provide a platform for sharing experiences and new learnings in the field of forest management.

13. Kerala Forest Research Institute (KFRI)

Kerala Forest Research Institute (KFRI) is founded in 1975 and incorporated in the Kerala State Council for Science technology and environment in the year 2002. The institute is located at Peechi, Thrissur of Kerala State.

Objectives and activities

- The institute is mandated to conduct research on tropical forest and forestry.
- The Institute is envisioned as a Centre of Excellence in Tropical Forestry
- It provides scientific support for decision making on matters related to forestry, with particular emphasis on conservation, sustainable utilization and scientific management of natural resources.
- The Institute undertakes consultancy programmes for national and International agencies on the issues relating to biological conservation and management of natural resources.
- KFRI is accredited as research centre of the FRI University Dehradun, University of Calicut and Cochin University of Science and Technology for enrolling students for research programme leading to award of PhD Degree.

14. National Biodiversity Authority (NBA)

The National Biodiversity Authority (NBA) was established by the Central Government in 2003 to implement India's Biological Diversity Act (2002) The

NBA with its headquarters in Chennai, Tamil Nadu, delivers its mandate through a structure that comprises of the Authority, Secretariat, SBBs, BMCs and Expert Committees.

Activities of NBA

The NBA advices central Government on matters relative to the conservation of biodiversity, sustainable use of biological resources and fair and equitable sharing of benefits arising out of the use of biological resources.

The NBA advices the state governments in the selection of areas of biodiversity importance to be notified under Sub-Section (1) of Section 37 as heritage sites and measures for the management of such heritage sites.

The NBA considers requests by granting approval or otherwise for undertaking any activity referred to in Sections 3, 4 and 6 of the Act.

Since its establishment, NBA has supported creation of SBBs in 29 States and facilitated establishment of around 62,500 BMCs.

15. National Board of Wildlife

National Board for Wild Life is a "Statutory Organization" constituted under the Wildlife Protection Act, 1972.

Activities and Functions of the Board

- The board is "advisory" in nature and advises the Central Government on framing policies and measures for conservation of wildlife in the country.
- It is a very important body because it serves as an apex body to review all wildlife-related matters and approve projects in and around national parks and sanctuaries.
- The Primary function of the Board is to promote the conservation and development of wildlife and forests.
- It has power to review all wildlife-related matters and approve projects in and around national parks and sanctuaries.
- No alternation of boundaries in national parks and wildlife sanctuaries can be done without approval of the NBWL.
- The NBWL is chaired by the Prime Minister. It has 47 members including the Prime Minister. Among these, 19 members are ex-officio members. Other members include three Members of Parliament (two from Lok Sabha and one from Rajya Sabha), five NGOs and 10 eminent ecologists, conservationists and environmentalists.

16. National Tiger Conservation Authority, New Delhi

The National Tiger Conservation Authority (NTCA) is a statutory body of the Ministry, with an overarching supervisory/coordination role, performing functions as provided in the Wildlife (Protection) Act, 1972.

Role and Objectives

It Provides central assistance to States under the ongoing Centrally Sponsored Scheme of Project Tiger for tiger reserves, for activities (recurring/non-recurring), as reflected in the Annual Plan of Operations of tiger reserves, based on their Tiger Conservation Plans is an important activity.

The NTCA/Project Tiger also conducts the country level assessment of the status of tiger, co-predators, prey and habitat once in four years, using the refined methodology, as approved by the Tiger Task Force.

- The NTCS Alerts the States as and when required and transmits backward / forward linkages of information relating to poachers
- It Advices the States for combing forest floor to check snares/traps and antipoaching operations
- Performing supervisory field visits through the National Tiger Conservation Authority and its regional offices
- Using information technology for improved surveillance (e-Eye system) through thermal cameras launched in Corbett
- Prepares a national database of individual tiger photo captures to establish linkage with body parts seized or dead tigers
- Assisting States to refine protection oriented monitoring through monitoring system for tiger's intensive protection and ecological status (M-STrIPES)

17. Wildlife Crime Control Bureau (WCCB)

Wildlife Crime Control Bureau is a statutory multi-disciplinary body established by the Government of India under the Ministry of Environment and Forests, to combat organized wildlife crime in the country. The Bureau has its headquarter in New Delhi and five regional offices at Delhi, Kolkata, Mumbai, Chennai and Jabalpur; three sub-regional offices at Guwahati, Amritsar and Cochin; and five border units at Ramanathapuram, Gorakhpur, Motihari, Nathula and Moreh.

Activities and Functions of the Bureau

- It is mandated to collect and collate intelligence related to organized wildlife crime activities and to disseminate the same to State and other enforcement agencies for immediate action so as to apprehend the criminals.

- It is mandated to establish a centralized wildlife crime data bank.
- It co-ordinates actions by various agencies in connection with the enforcement of the provisions of the Act.
- It assists foreign authorities and international organization concerned to facilitate co-ordination and universal action for wildlife crime control.
- It provides capacity building of the wildlife crime enforcement agencies for scientific and professional investigation into wildlife crimes and assist State Governments to ensure success in prosecutions related to wildlife crimes.
- It advices the Central Government on issues relating to wildlife crimes having national and international ramifications, relevant policy and laws.
- It also assists and advises the Customs authorities in inspection of the consignments of flora & fauna as per the provisions of Wild Life Protection Act, CITES and EXIM Policy governing such an item.

18. Zoological Survey of India (ZSI), Kolkata

The Zoological Survey of India (ZSI) was established on 1st July, 1916 to promote survey, exploration and research leading to the advancement in our knowledge of various aspects of exceptionally rich life of the erstwhile ′British Indian Empire′. The survey has its genesis in the establishment of the Zoological Section of the Indian Museum at Calcutta in 1875.

Objectives and Activities

- Exploration, Survey, Inventorying and Monitoring of faunal diversity in various States, Ecosystems and Protected areas of India.
- Taxonomic studies of all faunal components collected.
- Periodic review of the Status of Threatened and Endemic species.
- Preparation of Red Data Book, Fauna of India and Fauna of States.
- Bioecological studies on selected important communities/species.
- Preparation of databases for the recorded species of the country.
- Maintenance & Development of National Zoological Collections.
- Training, Capacity Building and Human Resource Development.
- Faunal Identification, Advisory services and Library Services.
- Publication of results including Fauna of India and Fauna of States.
- Environmental Impact Studies.
- Maintenance and Development of Museum at Headquarters and Regional Stations.

- Development of ENVIS and CITES Centers.
- Research Fellowship, Associateship and Emeritus Scientist Programmes.
- Collaborative research programmes on Biodiversity with other Organizations.
- GIS and Remote Sensing studies for animal diversity as well as for selected threatened species.
- Chromosomal Mapping and DNA finger printing

19. Wildlife Trust of India (WTI)

Wildlife Trust of India (WTI) is a leading Indian nature conservation organisation committed to the service of nature. Its mission is to conserve wildlife and its habitat and to work for the welfare of individual wild animals, in partnership with communities and governments. WTI's team of 150 dedicated professionals work towards achieving its vision of a secure natural heritage of India, in six priority landscapes, knit holistically together by nine key strategies or Big Ideas.

Role and objectives

- A secure natural heritage of India.
- To conserve wildlife and its habitat and to work for the welfare of individual wild animals, in partnership with communities and governments.

INTERNATIONAL ORGANIZATIONS

1. Center for International Forestry Research (CIFOR)

The Center for International Forestry Research (CIFOR), Indonesia is a non-profit scientific research organization established at Bogar, Indhonasia. It conducts research on the use and management of forests with a focus on tropical forests in developing countries.

CIFOR's research contributes to the three CGIAR System-Level Outcomes: reduced poverty, improved food and nutrition security, and improved natural resource systems and ecosystem services. It also contributes to the new United Nations Sustainable Development Goals and commitments made by countries in the Paris agreement and other accords related to the UN Framework Convention on Climate Change.

Role and Objectives

- **Forests and human well-being:** CIFOR aims to supply the evidence needed for policy decisions that enhance the contribution of forests to human well-being and prosperity and that make use of forests and their services and products to reduce poverty.

- **Sustainable landscapes and food:** CIFOR provides a broad perspective and landscape-scale comparisons across sites of the contributions that forests and tree-based agricultural systems make to healthy and diverse diets.
- **Equal opportunities, gender, justice and tenure:** CIFOR evaluates how the devolution of forest rights affects forests and conservation outcomes, livelihoods and local governance and thus informs decision-makers. CIFOR also seeks to advance gender equity and the empowerment of women and girls.
- **Climate change, energy and low-carbon development:** CIFOR's research aims to improve not only technical understanding of climate change and its interactions with forests and landscapes, but also understanding of the social implications of climate change, so that the interests of rural land users are reflected in decision making.
- **Value chains, finance and investment:** CIFOR aims to support various governance arrangements for achieving sustainability, along with inclusive business models that can improve benefit sharing, upgrade smallholder systems, and support responsible finance.
- **Forest management and restoration:** Research in this area addresses two issues: how rural people in developing countries can access forest resources and how forest resources can be managed more equitably while increasing forest production from multiple resources.

2. International Tropical Timber Organization (ITTO)

ITTO is an intergovernmental organization promoting the conservation and sustainable management, use and trade of tropical forest resources. Its members represent about 80% of the world's tropical forests and 90% of the global tropical timber trade.

The International Tropical Timber Organization (ITTO) was established under the auspices of the United Nations in 1986 amidst increasing worldwide concern for the fate of tropical forests. While almost everyone was alarmed at the rate of deforestation occurring in many tropical countries, there was also considerable agreement that the tropical timber trade was one of the keys to economic development in those same countries. The reconciliation of these two seemingly disparate phenomena is ITTO's story.

Role and Objectives

- ITTO provides an effective framework for consultation, international cooperation and policy development among all members with regard to all relevant aspects of the world timber economy.

- It contributes to the process of sustainable development.
- It enhances the capacity of members to implement a strategy for achieving exports of tropical timber and timber products from sustainably managed sources by the Year 2000.
- ITTO promotes the expansion and diversification of international trade in tropical timber from sustainable sources by improving the structural conditions in international markets.
- It promotes and support research and development with a view to improving forest management and efficiency of wood utilization as well as increasing the capacity to conserve and enhance other forest values in timber producing tropical forests.
- To develop and contribute towards mechanisms for the provision of new and additional financial resources and expertise needed to enhance the capacity of producing members to attain the objectives of this Agreement;
- It improves market intelligence with a view to ensuring greater transparency in the international timber market.
- ITTO promotes increased and further processing of tropical timber from sustainable sources in producing member countries with a view to promoting their industrialization and thereby increasing their employment opportunities and export earnings.
- ITTO encourages members to support and develop industrial tropical timber reforestation and forest management activities as well as rehabilitation of degraded forest land.
- ITTO improves marketing and distribution of tropical timber exports from sustainably managed sources.
- ITTO encourages information-sharing on the international timber market.

3. World Agroforestry Centre, Nairobi, Kenya

The World Agroforestry Centre (a brand name used by the International Centre for Research in Agroforestry, ICRAF), is an international institute headquartered in Nairobi, Kenya, and founded in 1978 as "International Council for Research in Agroforestry".

The Centre specializes in the sustainable management, protection and regulation of tropical rainforest and natural reserves. It is one of 15 agricultural research centres which makes up the global network known as the CGIAR.

The Centre conducts research in agroforestry, in partnership with national agricultural research systems with a view to developing more sustainable and productive land use. The focus of its research is countries/regions in the developing world, particular in the tropics of Central and South America, Southeast Asia, South Asia, West Africa, Eastern Africa and parts of central Africa.

In 2002, the Centre acquired the World Agroforestry Centre brand name, although International Centre for Research in Agroforestry remains its legal name and it continues to use the acronym ICRAF.

In 2017, ICRAF released a study at the UN Climate Change Conference that centers on Agroforestry and the emission of carbons from deforestation.

Role and objectives

- ICRAF Supports the development of national agroforestry policies
- It involves in development of tree varieties and strengthening of rural advisory service provisions
- It develops land health assessment methods and multiple scale geo-spatial analyses
- It Formulates systems science paradigms and support negotiation support tools
- ICRAF devices climate adoption approaches and develops bioenergy options.

4. International Union for Conservation of Nature (IUCN)

The International Union for Conservation of Nature (IUCN; officially International Union for Conservation of Nature and Natural Resources) is an international organization working in the field of nature conservation and sustainable use of natural resources.

It is involved in data gathering and analysis, research, field projects, advocacy, and education. IUCN's mission is to "influence, encourage and assist societies throughout the world to conserve nature and to ensure that any use of natural resources is equitable and ecologically sustainable".

Over the past decades, IUCN has widened its focus beyond conservation ecology and now incorporates issues related to sustainable development in its projects. Unlike many other international environmental organisations, IUCN does not itself aim to mobilize the public in support of nature conservation. It tries to influence the actions of governments, business and other stakeholders by providing information and advice, and through building partnerships. The organization is best known to the wider public for compiling and publishing the IUCN Red List of Threatened Species, which assesses the conservation status of species worldwide.

IUCN has a membership of over 1400 governmental and non-governmental organizations. Some 16,000 scientists and experts participate in the work of

IUCN commissions on a voluntary basis. It employs approximately 1000 full-time staff in more than 50 countries. Its headquarters are in Gland, Switzerland.

IUCN has observer and consultative status at the United Nations, and plays a role in the implementation of several international conventions on nature conservation and biodiversity. It was involved in establishing the World Wide Fund for Nature and the World Conservation Monitoring Centre. In the past, IUCN has been criticized for placing the interests of nature over those of indigenous peoples. In recent years, its closer relations with the business sector have caused controversy.

IUCN was established in 1948. It was previously called the International Union for the Protection of Nature (1948–1956) and the World Conservation Union (1990–2008).

5. World Wide Fund for Nature (WWF)

The World Wide Fund for Nature (WWF) is an international non-governmental organization founded in 1961, working in the field of the wilderness preservation, and the reduction of human impact on the environment. It was formerly named the World Wildlife Fund, which remains its official name in Canada and the United States. The Living Planet Report is published every two years by WWF since 1998; it is based on a Living Planet Index and ecological footprint calculation.

It is the world's largest conservation organization with over five million supporters worldwide, working in more than 100 countries, supporting around 1,300 conservation and environmental projects. WWF is a foundation, with 55% of funding from individuals and bequests, 19% from government sources (such as the World Bank, DFID, USAID) and 8% from corporations.

Role and objectives

- WWF involved in conserving the world's biological diversity
- It ensures that the use of renewable natural resources is sustainable
- It Promotes the reduction of pollution and wasteful consumption

6. International Union of Forest Research Organizations (IUFRO)

The International Union of Forest Research Organizations (IUFRO) is a non-profit, non-governmental international network of forest scientists, headquartered in Austria. The IUFRO unites more than 15,000 scientists in almost 700 Member Organizations in over 110 countries. Scientists cooperate in IUFRO on a voluntary basis.

It promotes global cooperation in forest-related research and enhances the understanding of the ecological, economic and social aspects of forests and trees. It disseminates scientific knowledge to stakeholders and decision-makers and contributes to forest policy and on-the-ground forest management. IUFRO's vision is of science-based sustainable management of the world's forest resources for economic, environmental and social benefits.

With the Strategy 2015–2019, IUFRO addresses five research themes and associated emphasis areas, and three institutional goals. The following five themes aim to guide the science collaboration within IUFRO's global network in the current Strategy period:

- Forests for People
- Forests and Climate Change
- Forest and Forest-based Products for a Greener Future
- Biodivesity, Ecosystem Services and Biological Invasions
- Forest, Soil and Water Interactions

The three institutional goals are:

- Goal 1 – Research Excellence: Strive for quality, relevance and synergies
- Goal 2 – Network Cooperation: Increase communication, visibility and outreach
- Goal 3 – Policy Impact: Provide analysis, insights and options

7. International Network for Bamboo and Rattan (INBAR)

The International Network for Bamboo and Rattan (INBAR) is an independent intergovernmental organization established in 1997 to develop and promote innovative solutions to poverty and environmental sustainability using bamboo and rattan. INBAR headquarters in China has regional offices in India, Ghana, Ethiopia, and Ecuador.

Role of INBAR

- End poverty in all its forms
- Provide affordable, sustainable and reliable modern energy services for all
- Access to adequate and affordable housing
- Efficient use of natural resources
- Address climate change
- Protect and restore terrestrial ecosystems.

Objectives

- **Policy shaping:**It provides affordable, sustainable and reliable modern energy services for all.

- **Representation and advocacy:** INBAR coordinates inputs on bamboo and rattan from a growing global network of members and partners, and to represent Members' needs in the global policy arena.
- **Knowledge sharing and learning:** INBAR exchanges and communicate lessons learned wideley, providing training and raising awareness of the relevance of bamboo and rattan as strategic resources and commodities that directly support a number of the Sustainable Development Goals, Aichi Targets, REDD+ and other global, regional or national frameworks.
- **Action research and country support:** INBAR promotes adaptive research and on-the-ground innovation by promoting pilot best-practice case studies, and supporting the upscaling of best practices and innovations across INBAR Member states and with others.

8. Forest Stewardship Council (FSC)

The Forest Stewardship Council (FSC) is an international non-profit, multi-stakeholder organization established in 1993 to promote responsible management of the world's forests. The FSC does this by setting standards on forest products, along with certifying and labeling them as eco-friendly.

Role and objectives

- The Forest Stewardship Council (FSC) shall promote environmentally appropriate, socially beneficial, and economically viable management of the world's forests.
- The world's forests meet the social, ecological, and economic rights and needs of the present generation without compromising those of future generations.

Goal

The FSC's stated mission is to "promote environmentally appropriate, socially beneficial and economically viable management of the world's forests". To this end the body has published a global strategy with five goals.

- Advancing globally responsible forest management.
- Ensure equitable access to the benefits of FSC systems.
- Ensure integrity, credibility and transparency of the FSC system.
- Create business value for products from FSC certified forests.
- Strengthen the global network to deliver on goals 1 through 4.

Areas of Activities

- Forests

- Chain of custody
- Social policy
- Monitoring and evaluation
- Quality assurance and
- Ecosystem services.

The FSC logo signifies that the product comes from responsible sources - environmentally appropriate, socially beneficial and economically viable. The FSC label is used on a wide range of timber and non-timber products from paper and furniture to medicine and jewelry and aims to give consumers the option of supporting responsible forestry.

9. Food and Agriculture Organization (FAO)

The Food and Agriculture Organization (FAO) is specialized agency of the United Nations that leads international efforts to defeat hunger. The headquarters is at Rome, Italy.

The goal is to achieve food security for all and make sure that people have regular access to enough high-quality food to lead active, healthy lives. With over 194 member states, FAO works in over 130 countries worldwide. We believe that everyone can play a part in ending hunger.

Role and objectives

1. Leverage global and national recognition of the importance of agricultural heritage systems and institutional support for their safeguard;
   - o Global recognition is obtained through the creation of the Agricultural Heritage Systems categories supported by governments, FAO governing bodies, UNESCO, World Heritage Centre and other partners:
   - o National recognition and awareness is raised by improving understanding of the threats that such agricultural systems face, of their global importance and of the benefits that they provide at all levels.
2. Capacity building of local farming communities and local and national institutions to conserve and manage GIAHS, generate income and add economic value to goods and services of such systems in a sustainable fashion;
   - o Identify ways to mitigate risks of erosion of biodiversity and traditional knowledge, land degradation and threats posed by globalization processes, and skewed policies and incentives;
   - o Strengthen conservation and sustainable use of biodiversity and natural resources, reducing vulnerability to climate change,

enhancing sustainable agriculture and rural development and as a result contributing to food security and poverty alleviation;

- o Enhancing the benefits derived by local populations from conservation and sustainable use of their resources and their ingenious systems and rewarding them through the payment for Environmental Services, Eco-labeling, Eco-tourism and other incentive mechanisms and market opportunities.

3. Promote enabling regulatory policies and incentive environments to support the conservation, evolutionary adaptation and viability of GIAHS;
   - o Assessment of existing policies and incentive mechanisms, and identification of modalities to provide support for sustainable agricultural practices.
   - o Promotion of national and international processes leading to improved policies and incentive mechanisms.

10. Intergovernmental Panel on Climate Change (IPCC)

The Intergovernmental Panel on Climate Change (IPCC) is a scientific and intergovernmental body under the auspices of the United Nations, set up at the request of member governments, dedicated to the task of providing the world with an objective, scientific view of climate change and its political and economic impacts.

- It was first established in 1988 by two United Nations organizations, the World Meteorological Organization (WMO) and the United Nations Environment Programme (UNEP), and later endorsed by the United Nations General Assembly through Resolution 43/53.
- Membership of the IPCC is open to all members of the WMO and UNEP. The IPCC produces reports that support the United Nations Framework Convention on Climate Change (UNFCCC), which is the main international treaty on climate change. The ultimate objective of the UNFCCC is to "stabilize greenhouse gas concentrations in the atmosphere at a level that would prevent dangerous anthropogenic [i.e., human-induced] interference with the climate system".
- IPCC reports cover "the scientific, technical and socio-economic information relevant to understanding the scientific basis of risk of human-induced climate change, its potential impacts and options for adaptation and mitigation."
- The IPCC was established to provide the decision-makers and others interested in climate change with an objective source of information about climate change.

■■■■

CHAPTER - 18

FORESTRY EVENTS / DATES

Forest as a biological resource has gained significant importance among several stake holders. This has resulted in establishment of important events and dates related to forest and environment. For each and every event and dates there is a theme and philosophy which needs to be disseminated across different sections of the society. Hence a chapter on Forestry Events / Dates has been designed and presented with the following sub headings.

i. Important Weeks related to Forestry

ii. Important Years related to Forestry

iii. Important Events / Dates related to Forestry

iv. Important Decades

I. IMPORTANT WEEKS

Event	Week	Significance
National Green Week	First week in February	• National Green Week is an annual event that empowers schools to engage in sustainability focused lessons, projects or activities between the first full week in February.
National Wildlife Week	First week of October	• The annual theme of the campaign is to promote the preservation of fauna. Overall goal of raising awareness to serve the long-term goal of safeguarding the lives of wildlife through critical action.
National Pollinator Week	Third week in June	• Pollinator Week has now grown into an international celebration of the valuable ecosystem services provided by bees, birds, butterflies, bats and beetles.
Recycle Week	20 to 26 June 2011	• The aim of the week is to encourage the public to recycle more, by demonstrating the benefits of recycling items from all around the home. Promoting kerbside, industrial and community

Event	Week	Significance
		recycling initiatives. Giving people the tools to minimise waste and manage material resources responsibly at home, work and school.
Zero Waste Week	First week of September	• Zero Waste Week is the annual awareness campaign for reducing landfill.

II. IMPORTANT YEARS

Event	Years	Significance
World Population Year	1974	• This is an year of commitment of the world community to work towards ensuring a better quality of life for man.
International Year of the Ocean (IYO)	1998	• International Year of the Ocean was adopted to focus and reinforce the attention of the public, governments and decision makers at large on the importance of the oceans and the marine environment as resources for sustainable development. • The major aim of the joint efforts during 1998 was to create awareness and obtain commitments from gove-rnments to take action, provide adequate resources and give the priority to the ocean and coastal areas which they deserve as finite economical assets.
International Year of Mountains (IYM)	2002	• The United Nations has designated 2002 as the International Year of Mountains (IYM) to raise awareness of the importance of mountains to the world and the necessity of preserving their fragile resources through sustainable development. • Throughout the year people all over the world will participate in events to celebrate mountains and discuss ways to promote their conservation and sustainable development. • IYM provides a unique opportunity to put mountain issues higher on the global agenda and should act as a springboard and catalyst for long-term, sustained and concrete action for mountains in the 21st Century. All countries have been encouraged to establish IYM committees and support national and local events which focus attention on policy and programs related to sustainable mountain development.
International Year of Ecotourism (IYE)	2002	• The IYE offers an opportunity to review ecotourism experiences worldwide, in order to consolidate tools and institutional frameworks that ensure its

Event	Years	Significance
		sustainable development in the future. This means maximizing the economic, environmental and social benefits from ecotourism, while avoiding its past shortcomings and negative impacts.
International Year of Freshwater (IYF)	2003	• The availability of clean, fresh water is one of the most important issues facing humanity today – and will be increasingly critical for the future, as growing demands outstrip supplies and pollution continues to contaminate rivers, lakes and streams.
International Year of Deserts and Desertification	2006	• The Year's Objective is to spread awareness about the desert areas of the world and especially the problem of desertification. This year attention is given - in addition to the world's deserts - to magnificent ecosystems, which have been home to ancient civilizations. The WDCD will therefore also celebrate the fragile beauty and unique heritage of our deserts.
International Year of Planet Earth	2008	• The United Nations General Assembly declared 2008 as the International Year of Planet Earth to increase awareness of the importance of Earth sciences for the advancement of sustainable development. • UNESCO was designated as the lead agency. The Year's activities spanned the three years 2006-2009. • The International Year of Planet Earth aims to ensure greater and more effective use by society of the knowledge accumulated by the world's 400,000 Earth scientists. The Year's ultimate goal of helping to build safer, healthier and wealthier societies around the globe is expressed in the Year's subtitle 'Earth science for Society'.
International Year of Sanitation	2008	• The progress towards achieving the sanitation MDG has been slow and varied in different parts of the world. The main objective in declaring 2008 as the International Year of Sanitation is to help get the sanitation MDG back on track for achieving the goal of halving the proportion of people without access to basic sanitation by 2015. • The year aimed to develop awareness and action to meet the Millennium Development Goals sanitation target. Particular concerns are: Removing the stigma around sanitation, so that the importance of sanitation can be more easily and publicly discussed. Highlighting the poverty reduction, health and other benefits that flow from better hygiene, household

Event	Years	Significance
		sanitation arrangements and wastewater treatment.
International Year of Natural Fibres	2009	• The IYNF was intended to cover a wide range of natural fibres, of animal and vegetable origin, and with uses ranging from luxurious apparel to traditional and non-traditional industrial uses. The IYNF was expected to raise the profile of such natural fibres. Observance of the Year was based on cooperation among producers of natural fibres to emphasise the positive qualities of natural fibres.
Year of the Gorilla	2009	• 2009 has been declared the Year of the Gorilla (YoG) in order to avoid the extinction of the largest living primate, by the UNEP Convention on Migratory Species (CMS), the UNEP/UNESCO Great Ape Survival Partnership (GRASP) and the World Association of Zoos and Aquariums (WAZA). This global campaign raised awareness and educated the public about gorillas and their threatened status, while at the same time raising funds for tangible on-the-ground conservation work.
International Year of Biodiversity	2010	• The United Nations declared 2010 to be the International Year of Biodiversity. It is a celebration of life on earth and of the value of biodiversity for our lives. The world is invited to take action in 2010 to safeguard the variety of life on earth: biodiversity. The United Nations proclaimed 2010 to be the International Year of Biodiversity, and people all over the world are working to safeguard this irreplaceable natural wealth and reduce biodiversity loss. This is vital for current and future human wellbeing. We need to do more.
International Year of Forests	2011	• To raise awareness and strengthen the sustainable management, conservation and sustainable development of all types of forests for the benefit of current and future generations.
International Year of Soils	2015	• The Food and Agriculture Organization of the United Nations has been nominated to implement the IYS 2015, within the framework of the Global Soil Partnership and in collaboration with Governments and the secretariat of the United Nations Convention to Combat Desertification. The IYS 2015 aims to increase awareness and understanding of the importance of soil for food security and essential ecosystem functions.

Event	Years	Significance
International Year of Pulses	2016	• An International Year designation provides an unprecedented opportunity to raise awareness and to celebrate the role of beans, chickpeas, lentils and other pulses in feeding the world. Even more importantly, it will be a galvanizing moment to draw together key actors to further the contributions pulses make to health, nutrition, and sustainability. • The IYP 2016 aims to heighten public awareness of the nutritional benefits of pulses as part of sustainable food production aimed towards food security and nutrition. The Year will create a unique opportunity to encourage connections throughout the food chain that would better utilize pulse-based proteins, further global production of pulses, better utilize crop rotations and address the challenges in the trade of pulses.
International Year of Sustainable tourism for all	2017	• The International Year aims to support a change in policies, business practices and consumer behavior towards a more sustainable tourism sector than can contribute effectively to the SDGs.

III. IMPORTANT DAYS

Months / Days	Events	Significance
February		
February 2	World Wetlands Day	• On this day, in 1971, the Ramsar Convention on Wetlands of International Importance was signed. Wetlands are a very important part of our biodiversity and it is essential to see that they are well protected.
February 27	International Polar Bear Day	• It is a day to learn more about the polar bear and conservation efforts where the polar bear is concerned.
February 28	National Science Day	• This day is celebrated to highlight the role of science in the protection of the environment.
March		
March 3	World Wildlife Day	• World Wildlife Day is celebrated to raise awareness on the conservation of wild fauna and flora coupled with the multitude of benefits provided by the wild flora and fauna.

Months / Days	Events	Significance
March 14	International Day of Action for Rivers	• The International Day of Action for Rivers is a day to celebrate victories such as dam removal and river restoration. • It is a day to take to the streets, demonstrate and demand improvements in the policies and practices of decision makers. • It is a day to educate one another about the threats facing our rivers, and learn about better water and energy solutions. • This day is born out of the first International Meeting of People Affected by Dams in Brazil in March 1997, and organised by the International Rivers NGO, the day has become global, with public events ranging from festivals, press conferences, kayaking events, river clean-ups, protests, religious gatherings and campaign launches.
March 20	World Sparrow Day	• World Sparrow Day is a day designated to raise awareness of the house sparrow and then other common birds to urban environments, and of threats to their populations. • It is an international initiative by the Nature Forever Society of India in collaboration with the Eco-Sys Action Foundation (France) and numerous other national and international organisations across the world.
March 21	World Forestry Day	• World Forest Day is celebrated on March 21st of every year to create awareness and the role of forests and need for its conservation.
March 22	World Water Day	• World Water Day is celebrated to create awareness on the conservation and judicial utilization of fast depleting water resources.
March 23	World Meteor-ological Day	• This day is celebrated to remind that weather is a integral part of the environment and to create awareness among the society.
April		
April 7	World Health Day	• The World Health Organization (WHO) was constituted on this day in 1948 and the Health Day is celebrated to indicate that health is important in the changing environment.

Months / Days	Events	Significance
April 22	World Earth Day	• In 1970 a group of people in the United States of America got together to draw the attention of the world to the problems being caused to the earth due to modernization. Since then this day has been celebrated all over the world as Earth Day.
May		
May 3	International Energy Day	• The World Energy Day is celebrated to create awareness towards conservation, rationale utilization of energy to promote energy sustainability.
May 8	World Migratory Bird Day	• World Migratory Bird Day (WMBD) was initiated in 2006 and is an annual awareness-raising campaign highlighting the need for the conservation of migratory birds and their habitats. • It has a global outreach and is an effective tool to help raise global awareness of the threats faced by migratory birds, their ecological importance, and the need for international cooperation to conserve them.
May 11	National Technology Day	• The National Technology Day was launched towards recognizing the successful test of nuclear tests at Pokhran. Also first, indigenous aircraft "Hansa-3" was test flown at Bangalore on this day and India also performed successful test firing of the Trishul missile on the same day. Considering above technological achievements on a particular date i.e. 11th May, the day of 11th May was chosen to be commemorated as National Technology Day. • This year the National Technology Day is marked as "Technology for Inclusive and Sustainable Growth".
May 14	Endemic Bird Day	• Endemic Bird Day is an annual event in India in May, coinciding with the Global Big Day, a worldwide effort to document as many species as possible in a single 24-hour period.
May 22	World Biodiversity Day	• The United Nations has proclaimed May 22 The International Day for Biological Diversity (IDB) to increase understanding and awareness of biodiversity issues.

Months / Days	Events	Significance
May 23	World Turtle Day	• The purpose of World Turtle Day, May 23, sponsored yearly since 2000 by American Tortoise Rescue, is to bring attention to, and increase knowledge of and respect for, turtles and tortoises, and encourage human action to help them survive and thrive
June		
June 5	World Environment Day	• On this day, in 1972, the Stockholm Conference on Human Environment was held in Sweden. There was a large gathering from all over the world and people expressed their concerns for the increasing environmental problems.
June 8	World Ocean Day	• The day is marked in a variety of ways, including launching new campaigns and initiatives, special events at aquariums and zoos, outdoor explorations, aquatic and beach cleanups, educational and conservation action programs, art contests, film festivals, and sustainable seafood events.
June 9	Coral Triangle Day	• The intention of Coral Triangle Day is to carry out the message of ocean conservation under the overall banner of the Coral Triangle Day. The Coral Triangle Day is an annual, open-sourced event that brings together individuals, organizations, and establishments on one special day of the year to shed light on ocean conservation and the numerous ways to protect and conserve the Coral Triangle.
June 15	Global Wind Day	• It is a worldwide event that occurs annually on 15 June. It is a day for discovering wind, its power and the possibilities it holds to change the world. It is a day when wind energy is celebrated, information is exchanged and adults and children find out about wind energy, its power and the possibilities it holds to change the world.
June 17	World Day to Combat Desertification & Drought	• This day is to create awareness on drought and desertification, its implications on society and ways to minimize the problem among the students.
July		
July 1 – July 7	Van Mahotsav Saptah	• Students may be encouraged to share on importance of this vanamohatsav week, its history and let them do some exercise on biodiversity & plant as many as tree.

Months / Days	Events	Significance
July 3	World Seabird Day	• The World Seabird Union, an organization comprised of the world's seabird societies is pleased to announce the first annual World Seabird Day on July 3rd, 2017. The last known sighting of the Great Auk (Pinguinusimpennis) was on July 3rd, 1844, and seabird scientists, managers and enthusiasts from around the world will be commemorating this date. Seabirds are 'canaries in the coal mine' for ocean health and the World Seabird Union aims to highlight the threats and challenges they face.
July 11	World Pollution Day	• World Population Day aims to increase people's awareness on various population issues such as the importance of family planning, including gender equality, poverty, maternal health and human rights. Let the students share their information through competitions.
July 26	International Mangrove Day	• This day is celebrated to create awareness towards conservation of Mangrove ecosystem.
July 29	International Tiger Day	• Global Tiger Day, often called International Tiger Day, is an annual celebration to raise awareness for tiger conservation, held annually on 29 July. It was created in 2010 at the Saint Petersburg Tiger Summit. The goal of the day is to promote a global system for protecting the natural habitats of tigers and to raise public awareness and support for tiger conservation issues.
August		
August 10	World Lion Day	• World Lion Day comes on August 10 of every year. World Lion Day aims to raise awareness of the majestic hunter, simply known as the king of the jungle. The Lions are one of the big cats in the Panthera genus and a member of the Felidae family. Lions in the wild exhibit very similar behavior, especially when smaller females or younger animals join up with the more dominant lions in the pride.
August 12	World Elephant Day	• The goal of International Elephant Day is to create awareness of the urgent plight of African and Asian elephants, and to share knowledge and positive solutions for the better care and

Months / Days	Events	Significance
		management of captive and wild elephants.
August 22	Honey Bee Day	• National Honey Bee Day (formerly National Honey Bee Awareness Day) is an awareness day started by beekeepers in the United States to build community awareness of the bee industry, through education and promotion.
September		
Septe-mber 16	World Ozone Day	• The United Nations declared this day as the International Day for the Preservation of the Ozone Layer. It is the day the Montreal Protocol was signed.
Septe-mber 21	Zero Emissions Day	• ZeDay has become a worldwide movement, and is observed annually, often with a theme, anything from "Buddy Up" (sharing ZeDay with friends and family) to "Get Creative".
Septem-ber 26	World Environmental Health Day	• World Environmental Health Day is annually observed in many countries around the world on September 26. The observance was inaugurated in 2011 in Indonesia. • World Environmental Health Day was launched by the IFEH Council (International Federation of Environmental Health). This organization dissem-inates knowledge about environmental health among people and promotes cooperation between countries to improve the environmental situation of our planet.
Sept-ember 28	Green Consumer Day	• The problems of consumerism and its impact on the environment is an area of major concern in today's world. Awareness building on the importance of recycling-reusing-reducing should be taken up seriously.
October		
October 3	World Nature Day, World Habitat Day	• The earth is the habitat of not only human beings but also all living creatures. Increasing human activities is threatening the habitat of other living things. Hence the days are celebrated
October 4	World Animal Day	• This day is celebrated to indicate that the welfare of animals has to be looked into and given due importance. .

Months / Days	Events	Significance
October 6	World Wildlife Day	• World Wildlife Day is an opportunity to celebrate the many beautiful and varied forms of wild fauna and flora and to raise awareness of the multitude of benefits that conservation provides to people. At the same time, the Day reminds us of the urgent need to step up the fight against wildlife crime, which has wide-ranging economic, environmental and social impacts.
October 13	International Day for Natural Disaster Reduction	• Due to a change in the environment there has been an increase in the number of natural disasters. Efforts have to be taken to reduce these disasters and hence the day is celebrated.
October 24	International Day of Climate Action	• International Day of Climate Action is an environmental commemorative day for raising awareness among people for organizing climate change actions around the world.
November		
Nove-mber 6	International Day for Preventing the Exploitation of the Environment in War and Armed Conflict	• The United Nations attaches great importance to ensuring that action on the environment is part of conflict prevention, peacekeeping and peace building strategies - because there can be no durable peace if the natural resources that sustain livelihoods and ecosystems are destroyed. On 27 May 2016, the United Nations Environment Assembly adopted resolution UNEP/EA.2/Res.15, which recognized the role of healthy ecosystems and sustainably managed resources in reducing the risk of armed conflict, and reaffirmed its strong commitment to the full implementation of the Sustainable Development Goals listed in General Assembly resolution 70/1, entitled "Transforming our world: the 2030 Agenda for Sustainable Development". It aims to educate people about the damaging effects of war and armed conflict on the environment.
Novem-ber 12	World Birds Day	• The day is celebrated to help raise awareness around a specific theme," towards birds conservation.
Novem-ber 14	World Energy Conservation Day	• World Energy Conservation Day is celebrated to highlight the importance of energy consumption and its use in our day-to-day life, its scarcity and its impact on sustainability of global eco systems.

Months / Days	Events	Significance
		It focuses our concentration on significant issues facing the future of mankind with respect to energy. This Day serves to spell a sense of urgency on the issues involved.
December		
December 2	National Pollution Prevention Day	• This day shares information about types of pollution, about reduction, laws, local issues and finding solution from students. Promotes ecofriendly practices like ban of polythene, use of cycle or by walk, plantation, water conservation practices.
December 5	World Soil Day	• World Soil Day (WSD) is celebrated to focus attention on the importance of healthy soil and advocating for the sustainable management of soil resources. • An international day to celebrate Soil was recommended by the International Union of Soil Sciences (IUSS) in 2002. • World Soil Day 2017 activities aim to communicate messages on the importance of soil quality for food security, healthy ecosystems and human well-being.
December 11	International Mountain Day	• International Mountain Day is "observed every year with a different theme relevant to sustainable mountain development. • FAO is the U.N. organization mandated to lead observance of International Mountain Day. The theme for International Mountain Day 2010 was "Mountain minorities and indigenous peoples." • It aims to raise awareness about indigenous peoples and minorities who live in mountain environments and the relevance of their cultural heritage, traditions and customs.
December 14	National Energy Conservation Day	• It is celebrated to aware people about the importance of energy as well as saving or conserving the more energy by using less energy. The exact means of energy conservation is using less energy by avoiding the unnecessary uses of energy.

IV. IMPORTANT DECADES

Event	Year	Significance
International Drinking Water Decade, 1981-1990	1980s	• The First Water Decade was the ten-year period 1981 - 1990 designated by the United Nations to bring attention and support for clean water and sanitation worldwide. The decade focussed on safe water and sanitation for everybody by 1990.
International Decade for Natural Disaster Reduction	1990s	• Its basic objective was to decrease the loss of life, property destruction and social and economic disruption caused by natural disasters, such as earthquakes, tsunamis, floods, landslides, volcanic eruptions, droughts, locust infestations, and other disasters of natural origin. • To reduce through concerted international actions, especially in developing countries, loss of life, property damage and economic disruption caused by natural disasters such as earthquakes, windstorms, tsunamis, floods, landslides, volcanic eruptions, wildfires and other calamities of natural origin such as grasshopper and locust infestations. • The basic idea behind this proclamation of the Decade was and still remains to be the unacceptable and rising levels of losses which disasters continue to incur on the one hand, and the existence, on the other hand, of a wealth of scientific and engineering know-how which could be effectively used to reduce losses resulting from disasters.
United Nations Decade of Education for Sustainable Development	2005-2014	• The basic vision of the DESD is where everyone has the opportunity to benefit from education and learn the values, behaviour and lifestyles required for a sustainable future and for positive societal transformation. • The primary goal for the DESD is laid out in the United Nations General Assembly resolutions 59/237 in which the General Assembly "encourages Governments to consider the inclusion of measures to implement the Decade in their respective education systems and strategies and, where appropriate, national development plans". • The General Assembly "invites Governments to promote public awareness of and wider participation in the Decade, inter alia, through cooperation with and initiatives engaging civil society and other relevant stakeholders, especially at the beginning of the Decade".
Water for Life Decade	2005-2015	• The goal during the Water for Life Decade was to promote efforts to fulfill international commitments in the water sphere by 2015. It is to raise the profile of water in the global agenda, and to focus the world's attention on the

Event	Year	Significance
		groundbreaking, lifesaving, empowering work done by those implementing water programs and projects. We've tried to be a bridge, to further cooperation between governments and other stakeholders, between nations and diverse communities, between economic interests and the needs of ecosystems and the poor. • To promote efforts to ensure the participation of women in water and sanitation.
United Nations Decade for Deserts and the Fight against Deserti-fication	2010-2020	• The Resolution states that the main goal of the Decade should be a greater focus on water-related issues at all levels and on the implementation of water-related programmes in order to achieve internationally agreed upon water-related goals contained in Agenda 21, the United Nations Millennium Development Goals and the Johannesburg Plan of Implementation. The motivation for this resolution was the Parties' concern about the deteriorating situation of desertification in all regions, which has far-reaching implications for the attainment of the Millennium Development Goals, particularly the eradication of poverty and ensuring environmental sustainability.
United Nations Decade on Biodiversity	2011-2020	• The UN Decade on Biodiversity serves to support and promote implementation of the objectives of the Strategic Plan for Biodiversity and the Aichi Biodiversity Targets, with the goal of significantly reducing biodiversity loss. • The goal of the United Nations Decade on Biodiversity is to support the implementation of the Strategic Plan for Biodiversity and to promote its overall vision of living in harmony with nature. • Provide a flexible framework for implementation of the Strategic Plan for Biodiversity 2011-2020 and the Aichi Biodiversity Targets. • Support and provide guidance to regional and international organizations with a role in implementation of the Strategic Plan and achievement of the Targets. • Continue to build public awareness of biodiversity issues.

CHAPTER - 19

ACTS AND POLICIES

The Forests in India are managed with strong policy directions along with suitable laws which facilitate forest administration and management. The country has witnessed wide range of policies and acts to protect, conserve, and manage resources. But these policies and acts are periodically changed or amended to suit the current demands. Hence the chapter on Acts and Policies are incorporated and presented with the following subtitiles.

- National Forest Policies – 1894, 1952, 1988
- Salient features
- Important Acts in Indian Forests
  - IFA,1927 and its amendments
  - WPA, 1972 and its amendments
  - FCA, 1980 and its amendments
  - EPA, 1986 and its amendments
  - Biodiversity act, 2002
  - Tribal act, 2006

I. FOREST POLICY - 1894

Background

Dr. Voelcker (1893) "Improvement in Indian Agriculture" Separate chapter for Forestry – Chap-8 Stressed formulating Forest policy of 1894

Salient Features

- The main object of the management of forests is public benefit
- Preservation of the climatic and physical conditions of the country
- Supplying valuable timber for commercial purposes
- Minor forests

- Pasture lands
- Permanent cultivation
- Fulfilment of the basic forest needs of the local people
- Maximum revenue

Defects in the National Forest Policy of 1894

- This policy was enunciated by the Britishers which was formulated to suit the needs of British regime.
- The policy has mandated public benefit as a sole issue and did not consider other conservation issues.
- This policy has indicated conversion of forest land for agriculture purposes which resulted in massive conversion of forests for agriculture purposes.
- This policy did not taken into consideration the growing demands of society and the industry which lead to increased deforestation.
- The policy did not taken into account the professional man power needs of the country through systematic forestry education.
- It was also silent on systematic research on forestry programme for production as well as conservation.
- Wildlife is a major part in forest management and the policy did not make any provision for conservation and management of wildlife population which have started decreasing due to more human interaction.
- The policy did not consider the regulation of water and other ecological resources which resulted in rapid loss of these resources.

II. NATIONAL FOREST POLICY -1952

Background

- The need for forest products and land for growing food because of increase in human and animal population
- Many industries utilising forest products
- No provisions with regard to the management of private forest
- It was considered the problem of shifting cultivation
- Importance to wildlife management
- Sustained yield of forest management

- Proportion of the forest cover with respect to the total geography area of the country.
- It forestry research and training gets significance.

Needs

- Balanced and complementary use of forest land
- Checking of, denudation in mountain region and erosion of treeless bank rivers
- Establishing tree lands for wherever possible (amelioration purpose)
- To allow the supplies of grazing, small wood for agricultural implements and firewood
- Sustained supply of timber for defence, communication and industry
- Fulfilment of revenue needs.

Functional Classification of Forest

- **Protected forest** – preserved for physical and climatic considerations
- **National forest** – maintained and managed to the needs of defence, communication, industry
- **Village forest**- to provide grazing, firewood, and small timber for local requirements
- **Tree lands** – outside forest management for amelioration conditions of the country

Issues in the National Forest Policy of 1952

After the enunciation of National Forest Policy 1952, tremendous development have happened across the country in terms of utilization of wood and wood products, mushrooming of industries, diversion of forest land for non forestry purposes, development of infrastructure like roads and communication facilities, construction of dams and reservoirs etc. which resulted in decline of forest cover at an alarming rate. It has been estimated that during the period between 1950s and 1980s the country has lost over 1.5 million ha per annum. This has necessitated a change in forest policy which has not been addressed by the National Forest Policy 1952.

i. **Production Vs Conservation:** The 1952 forest policy was aimed more towards production lead forestry programme which resulted in large scale deforestation. However during the period between 1960s and 1980s the emphasis was shifted from production forestry to conservation forestry and forestry had undergone a massive change.

ii. **Mushrooming of Industries:** After 1952 mushrooming of industries have been established across the country like saw mills, pulp and paper, plywood, veneer, particle board, match splints, sports goods, agriculture implements and construction industry. These industries have taken heavy toll of Indian forests and the 1952 forest policy have not directed any issues to meet the industrial raw material.

iii. **Government Vs. Participatory Management:** Traditionally Indian forests were managed by forest department with the stringent rules and regulations. The 1952 forest policy has not taken into consideration the role of people in managing the forests through a participatory mode which is a major drawback in the policy.

iv. **Maintenance of Ecological Balance:** The 1952 forest policy has not taken into consideration the maintenance of ecological balance. As indicated earlier due to continuous exploitation to meet domestic and industrial demand, the forests have been degraded which lead to imbalance in ecosystem. This has not been taken care of by the 1952 National Forest Policy.

v. **Rights and Concessions:** The 1952 National Forest Policy has not taken to consideration the Rights and Concessions of the local dwellers. Though it has given claims of neighbouring communities but there is no clarity on rights and concessions.

vi. **Forestry Education:** The 1952 forest policy dealt only with in service forestry education and does not taken into consideration development of professional forestry education.

vii. **Forestry Research:** Similarly to education the policy does not taken into consideration the research outside the forest department. It has not given any direction on agroforestry research which is one of the important land use systems to satisfy the demands of domestic and industrial requirement.

III. NATIONAL FOREST POLICY 1988

Background

The aim of the policy is to ensure environment stability and maintenance of ecological balance for the sustenance of all life forms, humans, animal and plant. This national forest policy is enunciated due to the following need:

- To meet the increasing energy demands of the populations
- To obey the new directive principles and duties given in the Indian constitution
- Constitutional amendments of 1976 and its stress on environment
- Naked growth of forest based industries in this era

- Need for the guidelines on conservation of wildlife

Basic objectives

- To maintain the environment stability and ecological restoration through preservation
- To check soil erosion and denudation in the catchment areas and mitigation of floods
- To checks the sand dune extension in desert and costal lands
- To increase the forest and tree cover in the demand and degraded lands
- To meet the requirement of the tribal peoples
- To conserve the natural heritage
- To increase the productivity of the forest to meet the various needs
- To increase the substitution of wood and reduction of wood usages
- To create a massive people movement and minimize the pressure on existing forest

Essential of forest management

- Existing forest should be protected and its productivity improved
- Diversion of agricultural lands to forestry should be discouraged
- Conservation of biodiversity should be strengthened through national park, sanctuary and biosphere reserves
- A forestation programme encouraged to meet the local requirement of fuel wood

Management strategy

- The area under forest should be a minimum of 1/3rd of the total geographical area in which hills and mountainous region should have 2/3rd cover
- Afforestation, social forestry and farm forestry programme are encouraged
- Rights and concessions of the local peoples are suitability addressed
- Diversion of forest lands for non-forestry purposes is discouraged
- Wildlife conservation is given more importance
- Shifting cultivation practices are discouraged and cultivation of perennial crops through horticulture and tree forming is encouraged
- Fire and grazing management is given special consideration

- Forest based industries are advised to meet their raw materials requirements from outside the forest
- Forest extension, research and personal management are more emphasized

Comparison of Forest policies

S.No	Parameters	Forest Policy of - 1894	National forest policy of 1952	National forest policy of 1988
1.	**Objectives**	Improvement of agriculture and public interest and rights	Balanced and complementary use of forest produce	Conservation of natural heritage, biodiversity and genetic resources
2.	**Forest Classification**	Forest for timber production Minor forests Preservation forests Pasture lands	Protection forests National forests Village forests Tree lands	No classification
3.	**Tribal people and forest**		Not much guidelines	Symbiotic relation-ship between forest department ' and tribal communities
4.	**Scope for increasing tree cover**		Increasing vegetative cover through planting on railways, PWD lands	Massive afforestation programme emphasized through various measures
5.	**Control of private forest**		Few guidelines given	Detailed guidelines given
6.	**Proportion of forest area**		1/3 rd of the area, 60% in hills and 20% in plains	33% under forests cover anc 66% tree cover in hills
7.	**Protection to wild life**		Protection of rare and endangered species only	Network of protection measures strengthened
8.	**Grazing**		Grazing is allowed with nominal fee	Grazing is allowed according to carrying capacity
9.	**Shifting cultivation**		No coercive measures to stop shifting cultivation	Discourage shifting cultivation

S.No	Parameters	Forest Policy of - 1894	National forest policy of 1952	National forest policy of 1988
10.	**Sustainable yield**		Limited scope	Detailed measures are given
11.	**Forest administration**		Trained foresters are recommended	Personal management for professional foresters
12.	**Forestry legislation**		Enactment of forest acts	Appropriate legislation is given
13.	**Forestry training education and research**		Few guidelines given through rangers college and universities	Detailed guidelines on training and scientific forestry research is given for increasing productivity, afforestation, wildlife and wood substitution
14.	**Good will of local peoples**		Instructs need for peoples cooperation	Focus on people's participation in forest management.

Important Acts in Indian Forest

i. Indian Forest Act, 1927 and its Amendments
ii. Wild Life Protection Act, 1972 and its amendments
iii. Forest Conservation Act, 1980 and its amendments
iv. Environment Protection Act, 1986 and its amendments
v. Biodiversity Act, 2002
vi. Tribal Act, 2006

I. Indian Forest Act, 1927 and its Amendments

The Indian Forest Act, 1927 was an act to consolidate the then existing laws relating to forest, the transit of forest products and duties that can be levied on "Forest Product" as defined in Section 2 (iv) (a) and (b) of the act. The act deals with reserved forest, village forest, protected forest, control over forests and lands not being property of government, the duty on timber and other forest produce, regulation transit of forest products, collection of timber, penalties and procedures, cattle – trespass, forest officers and other miscellaneous provisions.

This act does not lay down a specific definition for forests. The act establishes three categories of forests, reserve forest, protected forest and village forest. The reserved forests (section 3 to 27 of the Act of 1927) can be notified by the State Government on any forest land or waste land to which the government has ownership or right. To be categorized as a reserved forest, the land must be forest land or waste land in the absence of which the notification could be quashed. Section 26 of the Indian Forest Act 1927 prohibits a number of activities including making fresh clearings, tree felling, lopping, burning, grazing, quarrying, manufacturing activities, hunting, shooting, etc. in the forest. Violation of provisions of Section 26 specifically with regards to creating fire, felling, girdling, lopping, etc. of trees, quarrying and manufacturing operations or clearing breaking up of any land for cultivation is punishable with imprisonment for a term which may extend to two years or with fine which may extend to Rs. 20,000 but which shall not be less than Rs. 5,000.

For other offenses under Section 26 an imprisonment for a term which may extend to 6 months or with fine which may extend to Rs. 1,000 or with both and on the second and every subsequent conviction for the same offense, with imprisonment which may extend to 6 months or with fine which may extend to Rs. 2,000 or with both is prescribed. It also provides for adjudication of forest rights, uses by local people as allowed by forest officer, appeals and denotification. Section 25 specifically empowers the forest officer to stop any public or private way or water coarse in reserved forest. The act also provides for issuing notification regarding appointment of forest settlement officer to adjudicate rights and granting permissions for activities and claims.

Section 28 provides for assigning rights of reserved forests or protected forests or any forest belonging to the state government to any village community and provisions relating to reserved forest, protected forest or forest belonging to the government shall apply. Such forests are called village forests.

Section 29 provides for the notification of protected forests. Protected forests are also notified on forest land or waste land. The state government under provision of Section 30 may declare any tree or class of trees in a protected forest to be reserved, it can also declare any portion of a protected forest as closed for a term not exceeding 30 ears during which the rights of private persons shall be suspended provided that alternate rights are available in the remainder of the forest. It can also prohibit specified activities within the area Violation of prohibited activities in protected areas as prescribed in Section 30 and 32 are punishable offences liable to be punished with imprisonment for a term which may extend to 2 years or with a fine which may extend to Rs. 5,000 or with both and on the second and every subsequent conviction for the same offence, with imprisonment for a term which may extend to 2 years and with fine which may extend to Rs. 10,000.

The state government also has the powers to regulate or prohibit activities in any forest or waste land not being the property of the government after providing suitable opportunity to the owner of such forest or land.

Section	Amendment	Year	Reason for the Amendment
68. Power to compound offences. (1) The State Government may, by notification in the Official Gazette, empower a Forest officer (a) to accept from any person against whom a reasonable suspicion exists that he has committed any forest-offence, other than an offence specified in **section 62 or section 63**, a sum of money by way of compensation for the offence which such person is suspected to have committed, and (b) when any property has been seized as liable to confiscation, to release the same on payment of the value thereof as estimated by such officer. (2) On the payment of such sum of money, or such value, or both, as the case may be, to such officer, the suspected person, if in custody, shall be discharged, the property, if any seized shall be released, and no further proceedings shall be taken against such person or property. (3) A Forest-officer shall not be empowered under this section unless he is a Forest officer of a rank not inferior to that of a Ranger **and is in receipt of a monthly salary amounting to at least one hundred rupees**, and the sum of money accepted as compensation under clause (a) of sub-section (1) shall in no case exceed the sum **of fifty rupees.**	In sub-section (*3*) of section 68 of the Indian Forest Act, 1927, (*a*) The words "and is in receipt of a monthly salary amounting to at least one hundred rupees" shall be **omitted;** (*b*) For the words "fifty rupees", the words "ten thousand rupees" shall be **Substituted** (*c*) After sub-section (*3*), the following shall be inserted, namely: (*4*) In Scheduled Areas, the Forest-officer shall, before exercising the powers vested under this section, obtain and record the views of the Gram Sabha Concerned.	2012	(*i*) A Forest-officer referred to for the purpose of compounding of offence should not be inferior to that of a Ranger and is in receipt of a monthly salary amounting to at least one hundred rupees, and (*ii*) The sum of money accepted as compensation under clause (*a*) of subsection (*1*) shall in no case exceed the sum of fifty rupees. The monthly salary limit of a Forest Officer specified in sub-paragraph (*i*) of the preceding paragraph empowered to compound aforesaid offences has become redundant and it is therefore proposed to omit salary limit from aforesaid sub-section (*3*). The sum not exceeding rupees fifty accepted as compensation as mentioned in paragraph (*ii*) of the preceding paragraph fixed in 1927 is very less. It is also proposed to revise the said limit from fifty rupees to ten thousand rupees. It is also proposed to insert a new sub-section (*4*) in section 68 of the Indian Forest Act, 1927 so as to make it mandatory for the Forest-officer, to obtain and record the views of Gram

Section	Amendment	Year	Reason for the Amendment
			Sabha before compounding the offence so as to ensure the transparency in compounding of the offences and participation of the people at Gram Sabha level.
35. Protection of forests for special purposes. (1) The State Government may, by notification in the Official Gazette, regulate or prohibit in any forest or waste-land (a) the breaking up or clearing of land for cultivation; (b) the pasturing of cattle; or (c) the firing or clearing of the vegetation; when such regulation or prohibition appears necessary for any of the following purposes: (i) for protection against storms, winds, rolling stones, floods and avalanches; (ii) for the preservation of the soil on the ridges and slopes and in the valleys of hilly tracts, the prevention of land slips or of the formation of ravines, and torrents, or the protection of land against erosion, or the deposit thereon of sand, stones or gravel; (iii) for the maintenance of a water-supply in springs, rivers and tanks; (iv) for the protection of roads, bridges, railways and other lines of communication;	In the Indian Forest Act, 1927 (hereinafter referred to as the principal Act), in chapter V, for section 35 the following section shall be substituted, namely: **"35. The State Government shall not exercise any control over any forest or land not being the property of State Government except in accordance with law passed by Parliament for the purpose."** For section 36 of the principal Act, the following section shall be substituted, namely: **"36. The State Government shall frame rules for the purpose of carrying out the objects of this chapter."** **Section 37** of the principal Act shall be **omitted.** **Section 38** of the principal Act shall be **omitted**	**5th August 2016**	Indian Forest Act, 1927 provides for Reserved Forests, Village Forests and Protected Forests. However, Chapter V of the Act contain section 35 to 38 to regulate forest and lands which are not the property of the State Government, popularly called as 'Private Forests'. Some States Governments have started doing surveys to identify Private Forests. However, during the surveys carried, vast areas of plantations, agriculture, horticulture, garden lands have been included and identified as 'Private Forest'. In most of the cases, owners of land, which have been declared as Private Forests, are not aware that their lands, have been so declared. Before the start of these surveys no policy guidelines were announced by the State Governments, who were entrusted with the duty of doing survey.

Section	Amendment	Year	Reason for the Amendment
(v) For the preservation of the public health. (2) The State Government may, for any such purpose, construct at its own expense, in or upon any forest or waste-land, such work as it thinks fit. (3) No notification shall be made under sub-section (1) nor shall any work be begun under sub-section (2), until after the issue of a notice to the owner of such forest or land calling on him to show cause, within a reasonable period to be specified in such notice, why such notification should not be made or work constructed, as the case may be, and until his objections, if any, and any evidence he may produce in support of the same, have been heard by an officer duly appointed in that behalf and have been considered by the State Government. **36. Power to assume management of forests.** (1) In case of neglect of, or willful disobedience to, any regulation or prohibition under section 35, or if the purposes of any work to be-constructed under that section so require, the State Government may, after notice in writing to the owner of such forest or land and after considering his objections, if any, place the same under the control of a Forest-officer, and may declare that all or any of the provisions of this Act relating to reserved forests shall apply to such			In a number of cases, the land which is covered by 'Private Forests' is a mixed land where agriculture, horticulture, plantations, garden lands exists side by side. Hence, there is an urgent need to enact an exhaustive legislation on forests combining the Indian Forest Act, 1927 and the Forests (conservation) Act, 1980 and all other related Acts. However, in any case, no land should be declared as 'Private Forests', without enacting any legislation on the subject as, the present practice of notifying Private Forests, amounts to criminal trespass, which should be halted.

Section	Amendment	Year	Reason for the Amendment
forest or land. (2) The net profits, if any, arising from the management of such forest or land shall be paid to the said owner. **37. Expropriation of forests in certain cases.** (1) In any case under this Chapter in which the State Government considers that, in lieu of placing the forest or land under the control of a Forest-Officer, the same should be acquired for public purposes, the State 3overnment may proceed to acquire it in the manner provided by the Land Acquisition Act, 1894 (1 of 1894). (2) The owner of any forest or land comprised in any notification under section 35 may, t any time not less than three or more than twelve years from the date thereof, require that such forest or land shall be acquired for public purposes, and the State Government shall require such forest or land accordingly.			
2. Interpretation clause (7) "tree" includes palms, bamboos, skumps, brush-wood and canes.	In the Indian Forest Act, 1927, in section 2, in clause (*7*), the word "bamboos" shall be omitted. (*1*) The Indian Forest (Amendment) Ordinance, 2017 is hereby repealed.	23rd day of Novem ber, 2017.	"Tree", which includes palms, bamboos, stumps, brush-wood and canes. The bamboo, though taxonomically a grass is treated as tree for the purpose of the said Act, and therefore, attracts the requirement of permit for transit under the said Act. Although, many States have exempted felling

Section	Amendment	Year	Reason for the Amendment
	(*2*) Notwithstanding such repeal, anything done or any action taken under the Indian Forest Act, 1927, as amended by the said Ordinance, shall be deemed to have been done or taken under the corresponding provisions of the said Act, as amended by this Act.		and transit of various species of bamboos within the States, the inter-State movement of bamboos require permit when being in transit through other States. The farmers are facing hardships in getting the permits for felling and transit of bamboos within the State and also for outside the State, which has been identified as major impediment of the cultivation of bamboos by farmers on their land.

II. Wild Life Protection Act (1972) and its amendments

The Indian wild life act 1972 for the first time regulated the setting up and control of game parks to be referred to as National Parks and declared many species as protected animals and also provided for stringent punishments for poachers or other persons who killed wild animals. Effectively the act banned hunting for pleasure or sport

SALIENT FEATURES OF WILDLIFE ACT

- The wildlife act is aimed at preserving and protecting wildlife and came into effect in 1972.
- This act envisages national parks and wildlife sanctuaries as protected areas to conserve wildlife.
- Under this act, wildlife populations are regularly monitored and management strategies are formulated to protect them.
- The act covers the rights of forest dwellers. The act permits **restricted grazing in sanctuaries but prohibits the same in national parks.** The act also prohibits collection of non forest timber which might not harm the system.
- The rights of forest dwellers recognized by forest policy of 1988 were taken away by the amended wildlife act of 1991
- The act enables central government to enact the legislation.

- The act appoints a wildlife advisory board, wildlife warden and specifies their duties and powers.
- The act provides a comprehensive list of endangered species and prohibits hunting of the same.
- It protects endangered plants like Beddome Cycad, Blue Vandana, Ladies Sliper Orchid, Pitcher Plant, etc
- The act provides for setting up national parks, wildlife sanctuaries, etc.
- The act provides for constitution of central zoo authority
- There is provision for trade and commerce in some wild species with license for sale, purchase, transfer, etc
- The act imposes a ban on trade or commerce of commercial animals
- The act provides legal powers to officers to punish offenders.
- Under the act, captive breeding programs for endangered species have been initiated. Eg: Conservation project for Lion (1972), Tiger (1974), Crocodile (1974) & Brown antlered deer (1981)

The act has six schedules which cover the entire gamut of wild life.

- Schedule I and II are the most potent sections of the act. This section covers animals which are in the category of endangered species. The sections in this schedule give absolute protection to certain species and these cannot be infringed on any account. The value of these sections can bee seen from the fact that the famous actor Salman Khan was sentenced to 5 years rigorous imprisonment for shooting a black buck in Rajasthan. The case is under appeal in the high court. In addition 16 persons have been convicted and sentenced to various terms of prison up to 7 years for killing a tiger.
- Schedule III and IV. These also have roughly the same provisions of Section I and II, but cover animals that are not in danger of becoming extinct. The penalties under this section are also less than Schedule I and II.
- Schedule V delineates animals that can be hunted like ducks and deer's. For this purpose the hunter has to apply for a license to the District Forest Officer who will allow a hunter to shoot during a specific season and restricted area. Any infringement can lead to cancellation of the hunting license.
- Schedule VI concerns cultivation and plant life and gives teeth to setting up more protected animal parks.
- The Indian Wild life protection act is a god send for wild life conservation in India and gives sweeping powers to law enforcement authorities to punish anybody guilty under the act.

Amendments

Wildlife Protection Act, 1972

Section	Amendment	Year	Reason for the Amendment
Amendment of Long Title	In the Wild Life (Protection) Act, 1972 (hereinafter referred to as the principal Act), for the long title, the following long title shall be substituted	**2003**	An Act to provide for the protection of wild animals, birds and plants and for matters connected therewith or ancillary or incidental thereto with a view to ensuring the ecological and environmental security of the country."
section 2 clause: (1) "animal" includes amphibians, birds, mammals and reptiles and also includes, birds and reptiles clause: *(4)* The following clause shall be substituted, namely:- '*(4)* "Board" means a State Board for Wild Life constituted under subsection *(1)* of section 6; **Clause (8)** *:* Shall be omitted for clause *(9)*, the following clause shall be substituted, namely:- '*(9)* "Collector" means the chief officer-in-charge of the revenue administration of a district or any other officer not below the rank of a Deputy Collector as may be appointed by the State Government under section 18B in this behalf	"Animal" includes mammals, birds, reptiles, amphibians, fish, other chordates and invertebrates and also includes their young and eggs	**2003**	

Section	Amendment	Year	Reason for the Amendment
After Chapter IVA of the Wild Life (Protection) Act, 1972 (53 of 1972) (hereinafter referred to as the principal Act), the following Chapters shall be inserted, namely		**2006**	
Insertion of new Chapter IV D.	NATIONAL RHINO CONSERVATION AUTHORITY	**2017**	

III. FOREST CONSERVATION ACT (1980) AND ITS AMENDMENTS

An Act to provide for the conservation of forests and for matters connected therewith or ancillary or incidental thereto.

Be it enacted by Parliament in the Thirty-first Year of the Republic of India as follows:-

1. Short title, extent and commencement

(1) This Act may be called the Forest (Conservation) Act, 1980.

(2) It extends to the whole of India except the State of Jammu and Kashmir.

(3) It shall be deemed to have come into force on the 25th day of October, 1980.

2. Restriction on the dereservation of forests or use of forest land for non-forest purpose

Notwithstanding anything contained in any other law for the time being in force in a State, no State Government or other authority shall make, except with the prior approval of the Central Government, any order directing-

(i) that any reserved forest (within the meaning of the expression "reserved forest" in any law for the time being in force in that State) or any portion thereof, shall cease to be reserved;

(ii) that any forest land or any portion thereof may be used for any non-forest purpose;

(iii) that any forest land or any portion thereof may be assigned by way of lease or otherwise to any private person or to any authority, corporation, agency or any other organisation not owned, managed or controlled by Government;

(iv) that any forest land or any portion thereof may be cleared of trees which have grown naturally in that land or portion, for the purpose of using it for reafforestation.

Explanation - For the purpose of this section, "non-forest purpose" means the breaking up or clearing of any forest land or portion thereof for-

(a) the cultivation of tea, coffee, spices, rubber, palms, oil-bearing plants, horticultural crops or medicinal plants;

(b) any purpose other than reafforestation; but does not include any work relating or ancillary to conservation, development and management of forests and wildlife, namely, the establishment of check-posts, fire lines, wireless communications and construction of fencing, bridges and culverts, dams, waterholes, trench marks, boundary marks, pipelines or other like purposes.

3. Constitution of Advisory Committee

The Central Government may constitute a Committee consisting of such number of persons as h may deem fit to advise that Government with regard to-

(i) the grant of approval. under Section 2; and

(ii) any other matter connected with the conservation of forests which may be referred to h by the Central Government.

3A. Penalty for contravention of the provisions of the Act

Whoever contravenes or abets the contravention of any of the provisions of Section 2, shall be punishable with simple imprisonment for a period which may extend to fifteen days.

3B. Offences by the Authorities and Government Departments

(1) Where any offence under this Act has been committed -

(a) by any department of Government, the head of the department; or

(b) by any authority, every person who, at the time the offence was committed, was directly in charge of, and was responsible to, the authority for the conduct of the business of the authority as well as the authority; shall be deemed to be guilty of the offence and shall be liable to be proceeded against and punished accordingly.

Provided that nothing contained in this sub-section shall render the head of the department or any person referred to in clause (b), liable to any punishment if he proves that the offence was committed without his knowledge or that he exercised all due diligence to prevent the commission of such offence.

(2) Notwithstanding anything contained in sub-section (1), where an offence punishable under the Act has been committed by a department of Government or any authority referred to in clause (b) of sub-section (1) and it is proved that the offence has been committed with the consent or connivance of; or is attributable to any neglect on the part of any officer, other than the head of the department, or in the case of an authority, any person other than the persons referred to in clause (b) of sub-section (1), such officer or persons shall also be deemed to be guilty of that offence and shall be liable to be proceeded against and punished accordingly.

4. Power to make rules

(1) The Central Government may, by notification in the Official Gazette, makes rules for carrying out the provisions of this Act.

(2) Every rule made under this Act shall be laid, as soon as may be after it is made, before each House of Parliament, while it is in session, for a total period of thirty days which may be comprised in one session or in two or more successive sessions, and if, before the expiry of the session immediately following the session or the successive sessions aforesaid, both Houses agree in making any modification in the rule or both Houses agree that the rule should not be made, the rule shall thereafter have effect only in such modified form or be of no effect, as the case may be; so, however, that any such modification or annulment shall be without prejudice to the validity of anything previously done under that rule.

5. Repeal and saving

(1) The Forest (Conservation) Ordinance, 1980 is hereby replaced.

(2) Notwithstanding such repeal, anything done or any action taken under the provisions of the said Ordinance shall be deemed to have been done or taken under the corresponding provisions of this Act.

IV. ENVIRONMENT PROTECTION ACT (1986) AND ITS AMENDMENTS

This Act has been brought into force from November, 1986. Its salient features are:

- Conferring powers on the Central Government to:
  i. Take all necessary measures for **protecting** quality of **environment**,
  ii. Co-ordinate actions of States, officers and other authorities under this Act,

iii. Plan and execute a nationwide programme for prevention, control and abatement of environmental pollution,

iv. Lay down standards for discharge of environmental pollutants,

v. Empower any person to enter, inspect, take samples and test,

vi. Establish or recognise environmental laboratories,

vii. Appoint or recognise government analysts, lay down standards for quality of environment,

viii. Restrict areas in which any industries, operations or processes may not be carried out subject to certain safeguards,

ix. Lay down safeguards for prevention of accidents and take remedial measures in case of such accidents,

x. Lay down procedures and safeguards for handling hazardous substances,

xi. Constitute an authority for exercising powers,

xii. Issue directions to any person, officer or authority including the power to direct closure, prohibition or regulation of any industry, operation or process,

xiii. Require any person, officer or authority to furnish any prescribed information and

xiv. Delegate powers to any officer of a state or authority;

- It confers powers on persons to complain to courts regarding any violation of the provisions of the Act, after a notice of 60 days to the prescribed authorities;
- The Act makes it obligatory for the person in charge of a place to inform the prescribed authorities regarding any accidental discharge of any pollutant in excess of prescribed standards.
- The concerned authorities, on receipt of such information, shall take remedial measures to prevent or mitigate pollution caused by such accidents and expenses incurred by the authorities in respect of remedial measures are recoverable with interest from the polluter;
- It prescribes stringent penalties for violation of the provisions of the Act; and
- Jurisdiction of civil courts is barred under the Act.

A comprehensive Environment (Protection) Act came into being in 1986 to remedy the lacunae noticed in the earlier laws and to serve as a single legislation on the subject.

The Air (Prevention and Control of Pollution) Act, 1981 and the Water (Prevention and Control of Pollution) Act, 1974 were amended to bring their provisions at par with those of The Environment (Protection) Act, 1986 and to give more powers to the implementing agencies.

V. BIOLOGICAL DIVERSITY ACT, 2002

After an extensive and intensive consultation process involving the stakeholders, the Govt. of India has brought Biological Diversity Act, 2002.

- To regulate access to biological resources of the country equitable share in benefits arising out of the use of biological resources.
- To conserve and sustainable use of biological diversity.
- Setting up of National Biodiversity Authority (NBA), State Biodiversity Board (SBB) and Biodiversity Management Committee's. (BMC's).
- NBA and SBB are required to consult BMCs in decisions relating to bioresource / related knowledge within their Jurisdiction.
- To respect and protect knowledge of local communities traditional knowledge related to biodiversity.
- To secure sharing of benefits with local people as conservers of biological resources and holders of knowledge and information relating to the use of biological resources.
- All foreign nationals / organizations require prior approval of NBA for obtaining biological resources and / or associated knowledge for use.
- Indian scientists / individuals require approval of NBA for transferring results of research to foreign nationals / organizations.
- Conservation and development of areas of importance from the standpoint of biological diversity by declaring them as biological diversity heritage sites.
- Protection and rehabilitation of threatened species.
- Involvement of institutions of State Government in the broad scheme of the implementation of the Biological Diversity Act through constitution of committees.
- Protect India's rich biodiversity and associated knowledge against their use by foreign individuals and organizations without sharing benefits arising out of such use and check Biopiracy.
- Indian Industry needs prior intimation to SBB to obtain bioresource. SBB has right to restrict if found to violate conservation and sustainable use and benefit sharing.
- Provisions for notifying heritage sites by State Government in consultation with local body.

- Creation of National, State and Local Biodiversity Fund and its use for conservation of biodiversity.
- Prior approval is needed from NBA for IPRs in any invention in India or outside India on Bioresource

VI. FOREST RIGHTS ACT, 2006

The Scheduled Tribes and other Traditional Forest Dwellers (Recognition of Forest Rights) Act, 2006, in short is called Forest Rights Act.

Salient features of the Act

- Extends to the whole of India except the State of Jammu and Kashmir [Sec. 1(2)]
- Forest dwelling Scheduled Tribes and other traditional forest dwellers on all forest lands shall have following occupation of forest rights: [Sec. 3(1))
    a. Right to hold and live in the forest land under the individual or common occupation for habitation or for self-cultivation for livelihood.
    b. Community rights such as nistar.
    c. Right of ownership access to collect, use and dispose of minor forest produce.
    d. Other community rights of uses or entitlements such as fish, other products of water bodies, grazing and traditional seasonal resource access of nomadic or pastoralist communities.
    e. Rights, including community tenures of habitat and habitation.
    f. Rights in or over disputed lands under any nomenclature in any state where claims are disputed.
    g. Rights for conversion of Pattas or leases or grants issued by any local authority or any State Govt, on forest lands to titles.
    h. Rights of settlement and conversion of all forest villages, old habitation, unsurveyed villages and other villages in forests, whether recorded, notified, or not, into revenue village.
    i. Right to protect, regenerate or conserve or manage any community forest resource which they have been traditionally protecting and conserving for sustainable use.
    j. Rights which are recognized under any state law or laws of any Autonomous District Council or Autonomous Regional Council or which are accepted as rights of tribal under any traditional or customary law of the concerned tribes of any state.

k. Right of access to biodiversity and community right to intellectual property and traditional knowledge related to biodiversity and cultural diversity.

l. Any other traditional right excluding the traditional right of hunting or trapping or extracting a part of the body of any species of wild animal.

m. Right to in situ rehabilitation including alternative land in cases where the Scheduled Tribes and other traditional forest dwellers have been illegally evicted or displaced from forest land of any description without receiving their legal entitlement to rehabilitation prior to the 13th of December 2006.

- Notwithstanding anything contained in the Forest (Conservation) Act, 1980, the union government shall provide for diversion of forest land for the following facilities managed by the government, which involve felling of trees not exceeding seventy five trees per hectare, namely, schools, dispensary or hospitals, anganwadis, fair price shops, electric and telecommunication lines, tanks and other minor water bodies, drinking water supply and water pipelines, water or rain water harvesting structures, minor irrigation canals, non-conventional source of energy, skill up gradation or vocational training centres, roads and community centres. [Sec. 3(2)].
- Provided that such diversion of forest land shall be allowed if it is less than one hectare in each case and is recommended by the gram sabha.
- The forest rights recognized under this act in critical wildlife habitats of National Parks and sanctuaries may subsequently be modified or resettled , provided that no forest right holders shall be resettled or have their rights in any manner affected for the purposes of creating inviolate areas for wildlife conservation. [Sec 4(2)].
- The recognition and vesting of forest rights under this Act to the forest dwelling scheduled tribes and to other traditional forest dwellers in relation to any state or union territory in respect of forest land and their habitat shall be subject to the condition that such scheduled tribes or tribal communities had occupied forest land before the 13th of December, 2005. [Sec. 4(3)].
- A right conferred under this Act shall be heritable but not alienable or transferable and shall be registered jointly in the name of both the spouses in the cases of married persons and in the name of single head in the case of a household headed by a single person and in the absence of a direct heir, the heritable right shall pass on to the next-of-kin. [Sec. 4(4)]

- The forest rights to hold and live in the forest land shall be restricted to the area under actual occupation on the date of commencement of this Act and shall in no case exceed an area of four hectares. [Sec.4(6)].
- The forest rights shall be conferred free of all encumbrances and procedural requirements, including clearance under the Forest (Conservation) Act, 1980, requirement of paying the 'net present value' and 'compensatory afforestation' for diversion of forest land, except those specified in this Act [Sec.4(7)].
- The forest rights recognized and vested under this Act shall include the right of land to forest dwelling scheduled tribes and other traditional forest dwellers who can establish that they were displaced from their dwelling and cultivation without land compensation due to state development interventions and where the land has not been used for the purposes for which it was acquired within five years of the said acquisition. [Sec.4(8)].
- The gram sabha shall be the authority to initiate the process for determining the nature and extent of forest right, within the local jurisdiction, by receiving claims, consolidating and verifying them and preparing a map delineating the area of each recommended claim. It will then pass a resolution and forward a copy of the same to the sub-divisional level committee. (Sec.6(l)].
- The state government shall constitute a sub-divisional level committee to examine the resolution passed by the gram sabha and prepare the record of forest rights and forward it through the sub-divisional officer to the district level committee for a final decision. [Sec.6(3)].
- The state government shall constitute a district level committee to consider and finally approve the record of forest rights prepared by the sub-divisional level committee. [Sec.6(5)].
- The decision of the district level committee on the record of Forest rights shall be final and binding. [Sec.6(6)].
- Where any authority or committee or officer or member of such authority or committee contravenes any provision of this Act or any rule made there under concerning recognition of forest rights, it, or they, shall be deemed to be guilty of an offence under this Act and shall be liable to be proceeded against and punished with fine, which may extend to one thousand rupees. [Sec. 7].
- No suit, prosecution or other legal proceeding shall lie against any officer or other employee of the central government or the state government for anything which is in good faith done or intended to be done by or under this Act. [Sec. 10(1)]
- The ministry of the central government dealing with tribal affairs or any officer or authority authorised by the central government in this

behalf shall be the nodal agency for the implementation of the provisions of this Act. [Sec. 11].

The benefits that will accrue from the Act

- The Act results in recognition of forest rights of forest dwelling scheduled tribes and other traditional forest dwellers over the forest land under their occupation and their habitat for self-cultivation of the land for their livelihood.
- They will have access to, use or dispose of, minor forest produc.
- They will not face the threat of eviction or removal from forest land under their occupation.
- They will be entitled to the benefits of various schemes of the Government after vesting of the clear cut title of land in their favour.
- Since the Gram Sabhas have been designated as the competent authority for initiating the process of determining the nature and extent of individual or community forest rights that may be given to the scheduled tribes and other traditional forest dwellers, this would empower the local communities in management of their natural resources in tunc with the provisions of the PESA Act, 1996.
- The recognition and vesting of forest rights in the forest dwelling scheduled tribes and other traditional forest dwellers also includes the responsibility of protection, conservation and regeneration of wild life, forests and biodiversity, the Act envisages registration of the title of the forest land jointly in the name of both the spouses, where married, and in the case of single person headed households, in the name of the single head. This would also benefit the women dwelling in the forests

Annexure I

ABBREVIATIONS

ACF	:	Assistant Conservator of Forest
AFPs	:	Agroforestry Products
AFS	:	Agroforestry Systems
ANIIDCO	:	Andaman and Nicobar Islands Integrated Development Corporation Limited
AWB	:	Animal Welfare Board
BCFT	:	British Council for Forestry Terminologies
BNHS	:	Bombay Natural History Society
BSI	:	Botanical Survey of India
CAFRI	:	Central Agroforestry Research Institute
CAMPA	:	Compensatory Afforestation Management and Planning Authority
CAR	:	Central African Republic
CF	:	Conservator of Forest
CHP	:	Combined Heat and Power
CIDA	:	Canadian International Development Agency
CITES	:	Convention on International Trade in Endangered Species of Wild Fauna and Flora
CMS	:	Convention on Migratory Species
COCs	:	Carbon-Offset Credits
CRCs	:	Carbon Reduction Credits
CRIDA	:	Central Research Institute for Dryland Agriculture
CSE	:	Centre for Science and Environment
CZA	:	Central Zoo Authority
DAP	:	Di Ammonium phosphate

DBH	:	Diameter at Breast Height
DCF	:	Deputy Conservator of Forest
DFO	:	District Forest Officer
EPA	:	Environment Protection Act
ETPs	:	Planting Entire Transplants
FAO	:	Food and Agricultural Organization
FCA	:	Forest Conservation Act
FCRI	:	Forest College and Research Institute
FD	:	Forest Department
FDCs	:	Forest Development Corporation
FPCs	:	Forest Protection Committees
FRA	:	Forest Right Act
FRI	:	Forest Research Institute
FSC	:	Forest Stewardship Council
FSI	:	Forest Survey of India
FSI	:	Forest Survey of India
FYM	:	Farm Yard Manure
GA	:	Gibberellic Acid
GBH	:	Girth at Breast Height
GHG	:	Green House Gas
GIM	:	Green India Mission
GRASP	:	Great Ape Survival Partnership
ICAR	:	Indian Council of Agricultural Research
ICFRE	:	Indian Council of Forest Research & Education
ICFRE	:	Indian Council of Forestry Research and Education
IDB	:	International Day for Biological Diversity
IFA	:	Indian Forest Act
IFEH	:	International Federation of Environmental Health
IFS	:	Indian Forest Service
IGNFA	:	Indira Gandhi National Forest Academy
IGNFA	:	Indira Gandhi National Forest Academy
IIFM	:	Indian Institute of Forest Management

INBAR	:	International Network for Bamboo and Rattan
IPCC	:	Inter-Governmental Panel on Climate Change
IPIRTI	:	Indian Plywood Industries Research and Training Institute
IUCN	:	International Union for Conservation of Nature
IUFRO	:	International Union of Forest Research Organizations
IUSS	:	International Union of Soil Sciences
IYE	:	International Year of Ecotourism
IYF	:	International Year of Freshwater
IYM	:	International Year of Mountains
IYO	:	International Year of the Ocean
JFM	:	Joint Forest Management
JFMC	:	Joint Forest Management Committees
KFRI	:	Kerala Forest Research Institute
KW	:	Kilo Watt
LR	:	Long Rotation
MAB	:	Men and Biosphere
MOEF	:	Ministry of Environment and Forest
MoEF	:	Ministry of Environment & Forests
MSL	:	Mean Sea Level
MW	:	Mega Watt
NAP	:	National Afforestation Programme
NAPCC	:	National Action Plan on Climate Change
NASA	:	National Aeronautics and Space Administration
NBA	:	National Biodiversity Authority
NBWL	:	National Board for Wild Life
NFTS	:	Nitrogen-Fixing Tree Species
NGOs	:	Non-governmental organizations
NGT	:	National Green Tribunal Act
NOAA	:	National Oceanic and Atmospheric Administration
NRCAF	:	National Research Centre for Agroforestry
NTCA	:	National Tiger Conservation Authority
NWDB	:	National Wasteland Development Board

NWFP	:	Non-Wood Forest Products
PCCF	:	Principal Chief Conservator of Forests
PISFR	:	Preinvestment Survey of Forest Resources
PPP	:	Public private partnership
PRA	:	Participatory Rural Appraisal
RFA	:	Reserved Forest Area
SIDA	:	Swedish International Development Agency
SOC	:	Soil Organic Carbon
SPB	:	Seshasayee Paper and Boards Limited
SR	:	Short Rotation
SRW	:	Small Round Wood
TBO’s	:	Tree Borne Oil Seeds
TNPL	:	Tamil Nadu Newsprint and Papers Limited
TOF	:	Trees Outside Forest
UNESCO	:	United Nations Educational, Scientific and Cultural Organization
USAID	:	United States Agency for International Development
VAs	:	Village Associations
WAZA	:	World Association of Zoos and Aquariums
WCCB	:	Wildlife Crime Control Bureau
WCCB	:	Wildlife Crime Control Bureau
WHO	:	World Health Organization
WII	:	Wildlife Institute of India
WMBD	:	World Migratory Bird Day
WPA	:	Wild Life Protection Act
WSD	:	World Soil Day
WTI	:	Wildlife Trust of India
WWF	:	World Wide Fund for Nature
ZSI	:	Zoological Survey of India

REFERENCES

Anon. 1980. Forest (Conservation) Act, 1980. Ministry of Environment and Forests. Government of India, New Delhi.

Anon. 2002. Biological Diversity Act, 2002. Ministry of Environment and Forests. Government of India, New Delhi.

Anon. 2006. Forest Rights Act, 2006. Ministry of Tribal Affairs. Government of India, New Delhi.

Anon. 1894. Forest Policy, 1894. Ministry of Environment and Forests. Government of India, New Delhi.

Anon. 1927. Indian Forest Act, 1927. Ministry of Environment and Forests. Government of India, New Delhi.

Anon. 1952. National Forest Policy, 1952. Ministry of Environment and Forests. Government of India, New Delhi.

Anon. 1972. Wildlife Protection Act, 1972. Ministry of Environment and Forests. Government of India, New Delhi.

Anon. 1986. The Environment (Protection) Act, 1986. Ministry of Environment and Forests. Government of India, New Delhi.

Anon. 1988. National Forest Policy, 1988. Ministry of Environment and Forests. Government of India, New Delhi.

Chaturvedi, A.N and L.S. Khanna 1982. Forest Mensuration. International Book Distributors. Dehra Dun. P. 418.

Chundawat, B.S and S.K. Goutam 1996. Text book of Agroforestry, Oxford & IBM Publishing Co. Pvt. Ltd., Calcutta. P.188.

Dwivdei, A.P. 2006. A Text Book of Silviculture. International Book Distributors. Dehradun. P.505.

FAO. 2015. Global Forestry Resources Assessment – Desk reference, P.253.

FAO. 2015b. Global Forestry Resources Assessment – How are the world's forests changing? Second edition. P.54.

FSI. 2015. India State of Forest Report. Vol 16. P.300.

FSI. 2017. India State of Forest Report. Vol 17. P.352.

Khanna, L.S. 1984. Principles and Practice of Silviculture, International Book Distributers Dehradun. P.473.

Krishnamoorthy, T. 1993. Minor Forest Products of India. Oxford and IBH publishing Co. Pvt. Ltd. New Delhi. P.645.

Kumar, V. 2011. Nursery and Plantation practices in Forestry. Scientific Publishers (India). P.547.

Lal, J.B. 1992. India Forests - Myth and Reality. Natraj Publishers. Dehradun. P.304.

Luna, R.K. 2005. Plantation trees. International Book Distributors, Dehradun. P.975.

Manikandan, K and S. Prabu 2016. Indian Forestry – A Breakthrough Approaches to Forest Service. Jain Brothers (New Delhi). P.638.

Mehta, T. 2008. A Hand book of Forest Utilization. International Book Distributors, Dehradun. P.299.

Negi, S.S. 1986. A Hand Book of Social Forestry. International Book Distributors. Dehradun. P.178.

Negi, S.S. 1986. Forest Policy and Law. International Book Distributors, Dehradun. P.261.

Negi, S.S. 1992. Minor Forest Products. International Book Distributors, Dehradun. P. 168.

Parthiban, K.T. and R. Seenivasan 2016. Forestry Technologies – A Complete Value Chain Approach. Published by Scientific Publisher, New Delhi. P. 629 (ISBN No.: 978-93-86102-60-7).

Parthiban, K.T., B. Palanikumaran, N. Krishnakumar, R. Thiruniraiselvan and N. Kanagaraj. 2018. FAQ's on Forestry – For IFS Aspirants - A Decade of Solved Papers (Vol-I), Pearson Publishers, Noida **(In Press).**

Parthiban, K.T., B. Palanikumaran, N. Krishnakumar, R. Thiruniraiselvan and N. Kanagaraj 2018. FAQ's on Forestry – For IFS Aspirants - A Decade of Solved Papers (Vol-II), Pearson Publishers, Noida **(In Press).**

Parthiban, K.T., R. Jude Sudhagar, C. Cinthia Fernandaz and K.K. Suresh 2017. Agroforestry Strategies for Climate Change Mitigation and Adaptation. Published by Jaya Publishing House, New Delhi. P: 419 (ISBN No: 978-93-86110-53-4).

Parthiban, K.T., R. Jude Sudhagar, S. Umesh Kanna, S. Vennila, I. Sekar and K. Baranidharan 2016. Forestry – A Subjective guide for IFS Aspirants. P.622.

Parthiban, K.T., R. Umarani, S. Umesh Kanna, I. Sekar, P. Rajendran and P. Durairasu 2014. Industrial Agroforestry Perspective and Prospectives. Scientific Publisher. Jodhpur. P.396 (ISBN: 978-81-7233-905-0).

Ramachandran Nair P.K. 2008. An Introduction to Agroforestry. Springer Publishers. P.499.

Surrendran, C., K.T. Parthiban, K. Vanangamudi, and Balaji S. 2000. Vegetative Propagation of Trees – Principles and Practices. TNAU Publication. P.154.

■■■■